TELEGRAPH MESSENGER BOYS

Telegraph Messenger Boys

Labor, Technology, and Geography, 1850–1950

Gregory J. Downey

Routledge
Taylor & Francis Group

LONDON AND NEW YORK

First published 2002 by Routledge

Published 2013 by Routledge
2 Park Square, Milton Park, Abingdon, Oxon OX14 4RN
711 Third Avenue, New York, NY 10017, USA

Routledge is an imprint of the Taylor & Francis Group, an informa business

Library of Congress Cataloging-in-Publication Data is available from the Library of Congress.

Telegraph Messenger Boys: Labor, Technology, and Geography, 1850–1950 / Gregory J. Downey
ISBN 0-415-93108-8 (hbk); 0-415-93109-6 (pbk)   978-0-415-93109-0 (pbk)

CONTENTS

ACKNOWLEDGMENTS

This book originated as a doctoral thesis completed in March 2000 within two separate departments at the Johns Hopkins University in Baltimore: the Department of History of Science, Medicine, and Technology (HSMT) and the Department of Geography and Environmental Engineering (DOGEE). I wish to thank not only my advisers in each department—David Harvey, Bill Leslie, and Erica Schoenberger—but also the other faculty and graduate students in each department who endured my many presentations on "why telegraph messenger boys will make a good dissertation." My thanks go out in particular to the Brewer's Art crowd in HSMT and the Living Wage campaigners in DOGEE, two groups from whom I learned a great deal (you know who you are).

The work of transforming a long-winded dissertation into a readable book was largely completed while I was funded as a Woodrow Wilson Postdoctoral Fellow in the Department of Geography at the University of Minnesota, Twin Cities, through their Humanities Institute. I would like to thank not only the Woodrow Wilson Foundation, but also my sponsors "on the ground," Dan Brewer of the Humanities Institute and John Adams in the Department of Geography. Again, countless suggestions from the Minnesota community of faculty and students helped make this a better work.

I could not have completed the research for this project without the friendly and professional assistance of the staff at several libraries and manuscript collections: the Archives Center at the Smithsonian Institution's National Museum of American History, the Library of Congress, the New York Board of Education Archives at Columbia University Teachers' College, the Museum of the City of New York, the New York Public Library, the New York Historical Society, the Hagley Museum and Library, the University of Minnesota Library System, the University of Wisconsin Library System, the Wisconsin Historical

Society, and the Milton S. Eisenhower Library at the Johns Hopkins University. In the spirit of calling attention to the hidden workers who enable high-tech information technologies, I wish to particularly thank the apparently superhuman staff of the Interlibrary Loan Services at the Eisenhower Library, who conquered space and time to track down countless rare texts for me, saving me significant travel and cost.

Many former telegraph workers and their families offered their personal help on this project, granting me interviews and sending me materials. This project couldn't have been completed without their gracious assistance: Kenneth P. Akins, Warren Bechtel, Frank P. Bergman, Ben Bowlen, Edward S. Brown, Harold L. Carraway, Emma Cortez, Harry Dahlin, Edward Falborn, Lewis J. Feucht, Earl T. Goldsworthy, Bill Hamm, Arche M. Hartley, Ed Holberton, John Hollansworth, William H. Jackson, Bill Josanne, Hartley D. McTavish, James E. Rose, Edward F. Sanger, Gilbert Verret, Douglas Warren, Gordon K. Welner, and Harold M. Wilson. Thanks also to the telegraph history enthusiasts (many of them on the web) who helped me find these folks in the first place: John Barrows, Jr., Ernie H. Duesterhoeft, Bill Dunbar, Jack Durkin, Ralph Frank, Edward Gable, Russell Kleinman, Lea Macalee, Gerry Maira, Neal McEwen, Tom O'Sullivan, Greg Raven, Roger Reinke, and Bob Voss.[1]

Parts of the research behind this book have previously been published and presented in earlier forms, and I wish to thank the editors and organizers for letting me participate in the ongoing academic conversation on technology, labor, and geography. Portions of this work were presented at the 1999 conferences of the Society for the Social Study of Science (4S), the History of Science Society, and the Society for the History of Technology (SHOT). Variations on this work have been published in the SHOT "Mercurians" newsletter *Antenna* (editor Pam Laird), the 4S series *Knowledge and Society* (editor Shirley Gorenstein), and the SHOT journal *Technology and Culture* (editor John Staudenmaier).[2] As for the present book, thanks should go out to my editors at Routledge, Brendan O'Malley and David McBride, for their patience and care with this new author. I would also like to thank David Nasaw, who took an early interest in my interdisciplinary manuscript for his historical series.

Special thanks go out, as always, to my wife, Julie, and my son, Henry, for supporting me through the substantial time and effort that this project demanded.

Finally, I would like to dedicate this book to my two grandfathers, John Downey (1917–2001) and Edmund Krunfus (1917–1992), who together with the support and labor of their wives, Maebelle and Melinda, forged good and useful lives in the twentieth-century American Midwest, serving their communities and providing for their families through the twin labors of moving information and producing space.

LIST OF TABLES

LIST OF FIGURES

ACA	American Communications Association
ADT	American District Telegraph Company
AFL	American Federation of Labor
A&P	Atlantic and Pacific Telegraph Company
ARTA	American Radio Telegraphists' Association
AT&T	American Telephone and Telegraph Company
AU	American Union Telegraph Company
AWUE	Association of Western Union Employees
BW	*Business Week*
CIO	Congress of Industrial Organizations
CLB	*Child Labor Bulletin*
CTJ	*Commercial Telegrapher's Journal*
CTU	Commercial Telegrapher's Union
D&D	*Dots and Dashes*
E World	*Electrical World*
IDCH	*International Directory of Company Histories*
JoT	*Journal of the Telegraph*
MT	Magnetic Telegraph Company
NCLC	National Child Labor Committee
NLRB	National Labor Relations Board
NYT	*New York Times*
O&E World	*Operator and Electrical World*
PT	Postal Telegraph and Cable Company
PT	*Postal Telegraph Magazine*
T Age	*Telegraph Age*
T World	*Telegraph World*

T&T Age	*Telegraph and Telephone Age*
WHS	Wisconsin Historical Society
WU	Western Union Telegraph Company
WUA	Western Union archive, National Museum of American History
WU News	*Western Union News*

CHAPTER ONE

WHY TELEGRAPH MESSENGER BOYS?

Engineers have devised remarkable improvements in mechanisms for the transmission and reception of telegrams; but no one has been able to develop anything better calculated to get a telegram delivered than a quick-witted, energetic, dependable boy.

—W. S. Fowler, Western Union messenger manager, 1924[1]

There is only one branch of its business in which the telegraph company finds it impossible to economize. That is in the collection and delivery of messages. The messenger boy is a problem beyond the graphic chart of the engineer.

—*Telegraph Age* editorial, 1923[2]

In 1938, Western Union Telegraph Company employee Edward C. Brower retired at age seventy after over half a century of service to the telegraph industry, an event warmly publicized by the company in local New York City papers. But unlike other celebrated telegraph "old-timers," Brower had been neither a skilled Morse operator nor an inventive tinkerer and entrepreneur. He was, instead, the "general supervisor of messenger equipment" for Western Union, meaning he simply supplied the uniforms and bicycles for the company's nationwide force of some fifteen thousand teen-aged messenger boys.[3]

Brower's own fifty-five-year history through the telegraph industry paralleled that of the messenger service itself. Born in 1868, he started out at age 15 as a subcontracted "district" messenger boy on Wall Street, carrying telegrams and urgent notes for stock and commodity brokers. A year later, Brower got a job as an office messenger for another large New York City telegraph partner, the Ohio Central Railroad. In 1887, at age nineteen, Brower became the night manager at the Broadway office of the American District Telegraph Company (ADT), the firm from which Western Union (WU) subcontracted most of its New York City messenger boys. Here he learned the techniques for managing and monitoring messengers that would win him promotion in only two years to ADT's messenger inspector, a job that was apparently part drill sergeant and part private eye. In 1916, Brower officially transferred to WU as the new superintendent of supplies, earning a final promotion in 1921 and remaining "the old clothes man" until his retirement.[4]

It is perhaps fitting that one of Brower's last official duties was to address the graduating class of Western Union's New York City messenger continuation school. Brower had been born only a few years before the electric "callbox," a small wall-mounted switch that customers used to request messenger boys, transformed the telegraph messenger job from a straightforward delivery task into an open-ended pickup task. His job as messenger inspector, recorded in scrapbooks he kept through the turn of the century, exposed him to both the best and the worst of the messenger service—from proud boys who were sent on urgent missions as far away as South Africa, to desperate boys who stole money and jewels under the guise of the blue messenger uniform. His career in both WU and ADT took him through the complex piece-wage and stock-ownership agreements that bound the national telegraph monopoly together with its local subcontracted labor provider. And Brower's retirement in 1938—the final chapter in a striking and mythic messenger career success story—came just as contemporary messenger boys, participating fully in national telegraph unions for the first time, won the right to be included in the nation's first 25¢-per-hour minimum wage law, nearly doubling their earnings—a far cry from the easily dismissed messenger strikes over uniforms and wages of Brower's inspector days in the early 1900s.[5]

Today, even triumphant "messenger-makes-good" stories like Brower's are rare in the history of the telegraph. The story of America's first electrical communications network has been told many times in heroic Zane Grey style since the first branch of that network was erected between Washington, D.C., and the nearby city of Baltimore in the mid-1840s. Early accounts described the genius of inventor Samuel Morse in creating a workable system of "communication at a distance" through wires and batteries, poles and crossarms, and dots and dashes. Later tales written after the Civil War recounted the telegraph's effects on commerce and government, and the profit-making potential of the network as evidenced by the rise of Western Union to "natural monopoly" power. The twentieth-century histories that followed in the wake of two world wars still marveled at the telegraph's importance, but also described the telegraph's "inevitable" decline in the face of "superior" technologies such as airmail and the telephone. And at the turn of the twenty-first century, with new digital packet-switching computer networks begging analogies to the dots and dashes of old, the telegraph continues to be reimagined as "the Victorian Internet."[6]

In most of these stories, telegraph messenger boys appear on the margins as colorful but unimportant characters. From the start of the first commercial telegraph line in 1845, young boys were employed to ferry handwritten messages into and out of the electrical telegraph system, to and from individual customers. About the same time as such messages became known as "telegrams," the boys started to become known as "messengers." In the aftermath of the Civil War, messengers were clothed in military-style uniforms; at the turn of the twentieth century, they were supplied with modern safety bicycles. Thus has the whimsical image of the messenger survived in popular culture, still available today for rent on videocassette in old movies from the 1930s and 1940s starring child actors like Mickey Rooney and Billy Benedict (in what was then an innovative Western Union "product placement strategy").[7]

But real messenger work had little in common with its Hollywood depiction. In 1901, a Philadelphia "Western Union boy," for example, actually would have worked for the same franchise that employed Brower in New York City, ADT (the forerunner of today's ADT security systems). He would have been one of over 150 ADT messengers in the city, among their 22 branch offices. At age fourteen (assuming he hadn't lied on his job application) he would have worked ten–hour days, seven days a week, waiting on a bench in back of the district office with five other boys for his turn at the next messenger call from any of the 250 or so electric call-boxes on his local circuit (out of the more than 5,000 call-boxes scattered throughout the city). This one messenger would have handled about twelve telegrams a day, both pick-ups and deliveries, receiving two cents for every call—barely enough money to cover his weekly bicycle and uniform rental fees, let alone his meal, maintenance, and mending costs. With other odd service jobs that paid by the hour (or the minute) his overall wage might come to a little over $2 per week—almost exactly what he would have been making at the same occupation in the same city a quarter-century before.[8]

That such work existed at the birth of electric communications is not surprising; but the fact that messenger work not only persisted, but thrived for more than a century from the 1850s to the 1950s—through a tumultuous period of urban, technological, cultural, social, political, and corporate upheaval in America often referred to as the heyday of "modernity"—poses many historical questions. Consider the simple quantitative data available from the U.S. Bureau of the Census on the number of telegraph messengers employed from 1870 to 1950 (see figure 1.1). Rather than declining with the invention of the telephone in the 1880s, or even with the increased use of the automobile in the

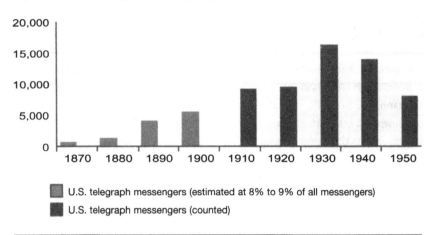

Figure 1.1. *U.S. telegraph messengers, 1870–1950*

U.S. telegraph messengers (estimated at 8% to 9% of all messengers)

U.S. telegraph messengers (counted)

Source: U.S. Census (1870–1950).
Note: Figures from 1870 to 1900 are conservative Census estimates based on 8% to 9% of the larger category of "messenger and errand and office boys" (the same portion of this larger category that telegraph messengers made up in 1910, when they were finally counted separately).

period from 1910 to 1930, messenger employment peaked around 1930 and only really began to drop off after World War II. As the epigrams heading this chapter suggest, such a pattern can be interpreted in many ways: Were messengers highly efficient workers who were hired in greater numbers because they could never be replaced even by "remarkable improvements in mechanisms," or were they an ever growing bottleneck in telegraphy, "a problem beyond the graphic chart of the engineer"?[9]

Quantitative sources like Census data only tell part of the story. Qualitative accounts in the popular media paint a more contradictory picture of these workers. In the late nineteenth century, the cartoon image of the slow, sleepy telegraph boy dominated, as newspaper editors and fellow operators alike struggled with the paradox of a "lightning-fast" information system that nevertheless seemed to rest on the speed of a lazy schoolboy. With the turn of the twentieth century, blaming the messenger shifted to blaming the messenger's environment, as a loose coalition of vice crusaders, child-savers, and education reformers battled to paint the messenger boy as a victim of the city, of the corporation, and of capitalism itself. With the increasing demand for the telegraph in the Roaring Twenties, the messenger's employers managed to transform him into the much-advertised image of the smiling, uniformed, industrial soldier, ready not only to deliver a holiday telegram with a smile, but competent to carry out product surveys, to deliver direct-marketing samples, and even to cover the office phones while the boss was away. And soon after World War II ended, a

Figure 1.2. *Three views of telegraph messengers*

A: Postal Telegraph advertisement, 1923. B: Cartoon of dozing messenger, 1906. C: Reformer photo of messenger, 1915.
Sources: *T&T Age* (November 1, 1923): i; *CTJ* (1906): 371; National Child Labor Committee, *Street-workers*, pamphlet no. 246 (New York: National Child Labor Committee, 1915), 8.

century after the first telegraph messenger appeared, finding any messenger image at all—save for a nostalgic one—was all but impossible[10] (see figure 1.2).

Together, the quantitative and qualitative evidence about the messengers suggests that their story is an important part of the history of so-called white-collar office work in the early twentieth century U.S. corporation. Telegraph messengers were merely the most public example of a widespread youth courier occupation that existed in other information networks such as the post office and the Bell System, not to mention the growing number of large private corporations that set up internal messenger services in "scientific" efforts to support the new "visible hand" of capitalist business management. Were these messengers lazy and slow, as derisive cartoons would suggest, or were they industrious and efficient, as advertisements declared? Were messengers exploited children, as the child-labor tracts argued, or up-and-coming young businessmen, as vocational guidance manuals suggested? And were messengers a throwback to preindustrial communications, as telegraph engineers lamented, or were they a textbook example of the crucial need for, and improved situation of, the skilled worker in the modern industrial age, as union leaders proclaimed?[11]

To answer such questions, this study takes what might be called an interdisciplinary or multidisciplinary approach, in two senses. First, it relies upon a diverse array of sources, both quantitative (like census records and wage reports) and qualitative (like personal recollections and literary works) as illustrated above. Traditional histories of the telegraph, focusing on communications technologies and their inventors, typically cite laboratory sketches, patent agreements, and engineering notebooks. But a study of telegraph labor must instead include hiring records, work rules, union contracts, management minutes and employee interviews. Studying such a popular labor force as the messenger boys brings popular advertisements, dime novels, and muckraking news reports into the mix as well. And to the degree that it is possible over a historical distance of more than a century, this study strives to let the messengers' own voices be heard.[12]

But besides new sources, a multidisciplinary history of telegraph messenger work demands a new interpretation of these sources, using several different analytical frameworks at once—after all, facts and figures, not to mention memories and moments, amount to little without some sort of context. This book draws on the interpretive traditions not only of history, but of geography as well, since in a study of a national communications network like the telegraph, not just technologies and laborers, but institutions and localities must also take the stage. Two theoretical frameworks, one drawn from the subfield of history of technology, and one drawn from the subfield of human geography, are crucial to understanding the arguments and the evidence about the telegraph messengers that are presented in this book: the idea of a "socially constructed technological system" and the concept of "socially produced space and time."[13]

Take the concept of a socially constructed technological system first. To conceive of a "technological system" is to treat individual technologies—be they physical devices or scripted procedures—not in isolation from each other, but together in the service of larger goals. In this view, individual inventors are less important than "system builders," those innovators and entrepreneurs who are able to combine particular technologies, sources of capital, and management expertise into cohesive institutions serving some wider purpose. For example, the various technologies that made up the telegraph network in America—senders and receivers, printers and repeaters, lines and poles, typewriters and pneumatic tubes, batteries and dynamos—each coevolved with the others, under the conscious control of communications providers and equipment manufacturers. But these aren't the only groups of people who shape technology. Other actors, such as government regulators or individual consumers, influence the evolution of technological systems as well. Thus such systems can be said to be socially shaped or socially constructed, taking both their form and the popular understanding of their purpose at any given moment from negotiations among these different groups of actors. Yet even socially constructed systems are still bound by the physical constraints of each individual technology's capacity to affect the material world.[14]

Information networks are inevitably socially constructed technological systems. This book argues that the technological network of the telegraph was more than just a combination of electromechanical systems; it was also a combination of systems of labor, in which messenger boys served different functions at different moments—sometimes working as technological components themselves, sometimes being sold as commodities along with the telegrams they carried, and sometimes acting as agents of change within the technological network itself. Messengers were not simply rendered obsolete by the slow and steady advance of technology—whether in telegraphy, telephony, or airmail. Instead, over the course of a century, they both cooperated in maintaining their usefulness to the telegraph, and fought to change their relationship to the telegraph in a way that would ultimately bring about their own exit from the industry.[15]

Like the notion of a socially constructed technological system, the idea of socially produced space and time sounds counterintuitive at first, but is actually a very useful theoretical tool. Both "space" and "time" are somewhat paradoxical concepts, for while they stand as abstract and absolute resources available in finite quantity, they are also part of a socially constructed reality, with different cultures in different times and places having entirely different conceptions of what it means to be "near" or "far," to move "quickly" or "slowly." Human cultures in different times and places have invented different concepts and technologies to alter their experiences of both space and time; thus, both space and time themselves can be thought of as commodities that are "produced." For example, in the early-twentieth-century telegraph, fitting more floors on a building produced more office space for operators and machinery; speeding up the machines that controlled telegraph transmission and reception produced more time for additional messages to be sent. But just as space and time are both physical and social phenomena, the production of space and time may have both physical and social implications as well, with space arranged specifically to exclude certain disempowered persons, or time arranged specifically to accommodate certain powerful persons. In all of these ways, the spatialities and temporalities that societies construct through their technologies—again, both physical artifacts and social practices—have real effects on how citizens live their lives.[16]

Information networks are inevitably involved in the social production of space and time. This book will argue that the telegraph industry produced certain spaces of control over information at both national and local scales. Besides producing wired "virtual" spaces for the transport of electrical information, the telegraph companies had to produce specific physical spaces for the transport of written information, spaces that were necessary to the messenger service. From numbered benches and uniform lockers in the back rooms and basements of telegraph offices where messengers waited out of public sight, to parade grounds, lecture halls, and "vice-free" business districts in the wider city where messengers were displayed before the public, the industry attempted to control the very urban space that grounded each local telegraph office.[17]

Taken together, these two theoretical frameworks from history and geography suggest that analyzing social relations is essential in trying to understand the production of human innovations, whether those innovations are technological systems or spatial/temporal patterns. To speak of the "social production" and "social construction" of innovation, then, is really to speak of the social relations of human labor. Many historians of the telegraph industry have investigated the telegraph operators—those laboring men and women who moved telegrams electrically, whether working alone in country offices at simple mechanical Morse keys (as was usually the case in the mid-nineteenth century), or working side-by-side with hundreds of others in noisy urban skyscrapers at expensive Simplex telegraph printers (as was more likely in the early twentieth century). Such studies have focused on a few key elements: unionization among these skilled office workers, the rise of a new category of "white-collar work" to describe such labor (with "white" often referring to the ethnicity of the workers as well), and the feminization of that white-collar work, meaning both the entry of more and more women into such occupations and the cultural redefinition of those occupations themselves as "women's work." Thus class, ethnicity, and gender analyses have all been successfully brought to bear on the problem of telegraph labor in the production of both telegraphic technologies and telegraphic spatio/temporalities.[18]

However, studying the telegraph messengers demands several additional tools as well, because messenger work was fundamentally distinct from operator work in both space and time. Telegraph managers, engineers, operators, and clerks—the classic actors in most business and technology histories—dealt with a time-pressure production floor of electromechanical senders, routers, repeaters, and receivers of all kinds. But telegraph messengers worked literally outside of the production floor, as service workers in the customer location, mediating between the customer and the rest of the telegraph network. Issues of class, ethnicity, gender, and also age became all the more important, as the ability of a messenger to move inconspicuously in different urban settings and to speak intelligibly to different groups of consumers made all the difference in getting the message through. Thus, while the telegraph network had the spatial and temporal characteristics of both a high-tech "white-collar" production industry and a low-tech "blue-collar" service industry, those characteristics were embodied hierarchically in different groups of employees. It is this tension that makes the history of the telegraph messengers an important topic of study.[19]

The following chapters carry these themes of labor, technology, and space-time through different aspects of messenger work at key moments of telegraph history, in roughly chronological order. Chapter 2 begins with a brief description of the telegraph network as it grew in the mid-nineteenth century from regional to national in scale, as a service moving information electrically between cities. Instead of focusing on particular telegraph inventions such as Morse sounders, automatic repeaters, and duplexed lines, the chapter concentrates on the spatial and temporal changes that such inventions enabled within the technological system of the telegraph. And instead of highlighting the few system builders who created the telegraph industry, the chapter spotlights the day-to-day laborers who made the telegraph industry work. The telegraph in the United States was unusual in that it was privatized under the all-but-monopoly control of Western Union managers, rather than nationalized under the government post office as in most other industrializing countries of the time; but private or public, the network was useless without a growing army of operators, clerks, and messengers. The overlapping maps of labor markets, customer markets, technological infrastructure, and institutional control resulted in an "uneven geography" of telegraph service in the young industry.[20]

Chapter 3 shifts from the "global" to the "local," describing how the telegraph grew not only as a system of electrical communication *between* cities, but as a system of written communication *within* cities. Again, even though this system was institutionalized under hundreds of private franchise agreements with district telegraph companies like ADT, rather than subcontracted to the civil-service post office, it was nevertheless enabled by a nationwide legion of young telegraph messengers, reflecting more the vibrant diversity of their local communities than the discriminatory monotony of their national employing agencies. Contrary to the usual story of competition between communications companies waged solely through inventors, technologies, and patents, this chapter illustrates the value of place-bound messenger labor to the national telegraph, revealing how the district messenger companies and the boys they employed became key resources in the industry's "Gilded Age" competitive battles.

Chapter 4 describes the physical movements of messenger boys through the spaces of telegraph production and consumption, from their wheeled freedom through the dangerous city streets to the precise discipline of their hidden urban offices. While operators were physically enmeshed in an electrical communications system that resembled a factory production floor, messengers extended this communications system to a system of transportation reaching all corners of the urban environment. Telegraph companies were fond of claiming that their communications technologies "annihilated space through time," but those companies were physically trapped within an urban system of buildings, roads, elevators, and corridors—technologies that, in bringing massive numbers of people together in social action simultaneously, "annihilated time through space." This built environment could be traversed only by messenger boys; but in doing so, messengers became the scapegoats for any delay in the "speed-of-light" telegraph, presenting an incongruous premodern figure that somehow didn't fit with the idea of "lightning wires" and Scientific Progress. Yet messengers kept in their heads complex virtual maps, not only of the urban landscape they traversed every day, but of the national landscape that demanded that telegrams of differing lengths and differing destinations be priced differently as well. Thus messengers occupied a key position in this information network at what might be considered the boundary between the virtual and the physical.[21]

Chapter 5 takes up the dual identity of messengers as both workers and commodities, analyzing them together with the most widely recognized product of the telegraph industry: the telegram. Not only did messengers participate in the "commodity fetishism" of the telegram, helping to obscure the very difficult labor that went into the telegram's production and transportation by their own playful and youthful appearance, but messengers were themselves commodified twice over—once as labor, selling their telegraph-handling skill to Western Union for a piece wage, and once as product, resold to consumers as "boys for every occasion" at an hourly rate in a relationship that might be considered the first modern "office temporary" role. This transformation of the messenger from labor into product occurred together with the transformation of the telegram itself—the "yellow blank," originally conceived as an all-purpose, general-audience information commodity in the nineteenth century, was in the twentieth century increasingly marketed either to businesses as a vehicle for coordinated, nationwide advertising messages, or to individuals as a heartfelt but practical social greeting. Paradoxically, this narrowing of the telegram's purpose both relied upon and repudiated the classic role of the messenger boy as a bearer of bad tidings. Thus, two ostensibly separate information technologies—messengers and telegrams—actually coevolved.[22]

Chapter 6 describes how the socioeconomic class of the messenger worked together with cultural understandings of gender and maturity—the assumptions and limits of masculinity and femininity, childhood and adulthood—both to broaden messenger duties and to limit the scope of their urban access. Telegraph managers chose young men as their messengers (as opposed to, say, adult men or young women) for very particular reasons. Messengers had to be low-wage and controllable, but diligent and trustworthy. They needed to be instantly recognizable, but also unobtrusively invisible. Messengers had to be able to access places and activities in the city that most urban women could not, but unable to demand the wages and respect of an urban man. Thus changes in the cultural meanings behind both sex and age were important aspects of "messenger technology," aspects illustrated well in the urban "vice" battles of the 1900s and 1910s, which cast messengers not as young apprentices of the high-technology world, but as economically exploited (and morally endangered) waifs. The resolution to these battles came not through altering the commodity of the telegram or the business practices of the telegraph industry, but instead through limiting the urban spatial and temporal access of the messenger.[23]

Chapter 7 explores how messengers affected the links between the three contemporary information networks of the telegraph, telephone, and post office from the 1870s to the 1930s. This chapter argues that a study of the messengers should not only be a study of the telegraph, but must connect to the story of the larger internetwork of competing and cooperating communications systems that the messengers negotiated in the course of their daily labor. The fact that a telegram sold by the telegraph network could actually be shepherded by messengers (or in defiance of messengers) through the other two networks on its way to the final consumer illustrates well that the three information networks of telegraph, telephone, and post office constituted a sort of multimodal information internetwork that began and ended with young boys but encompassed a variety of technologies, institutions, and geographies in between. Through the early part of the century, those technologies, institutions, and geographies of communication shifted considerably: as the telephone spread into the nation's homes and businesses, AT&T seized control of the telegraph during the early 1910s; as the post office grew in laborers but shrank in offices, the postmaster general seized control of both the telegraph and telephone during World War I; and even as the telegraph experimented with new technologies, its messenger force steadily grew. But through it all, in the continually-reproduced information internetwork of competing and cooperating technological systems, messengers and other laborers like them remained crucial "boundary workers."[24]

Chapter 8 takes the story into the Great Depression, discussing messenger career and education prospects in the 1920s and 1930s. Whether or not there still existed any real chance of career advancement for the messengers during this period, it was crucially important to the telegraph managers that the public *think* there was. A "myth of messenger advancement" had to be upheld at all costs: to public school officials, to child-labor reformers, to telegraph customers, and to the messengers themselves. By the 1920s, this myth had become so important that it attained a material expression in a quasi-public school run by Western Union in New York City. Thus the world's largest telegraph company found itself in the education business, in a striking example of the historical shift from apprentice training to "vocational education" that accompanied turn-of-the-century urbanization, immigration, and industrialization. Such were the social (and spatial) costs of continuing to employ the nation's single largest child-labor army of some fifty thousand boys per year.[25]

Chapter 9 recounts the story of messenger involvement with labor unions from the 1870s to the 1940s, to show how the pattern of messenger labor actions shifted from local, abrupt, and generally unsuccessful strikes to national, sustained, and ultimately successful organizations. Messengers were usually thought to be unskilled laborers by telegraph managers and operators alike, and generally came from poor or working-class families (especially in cities). The abstract Marxian class relation between worker and employer—the seller of labor power and the purchaser of it—was never more concrete than with the messengers, who remained subcontracted, piece-wage workers for nearly a century. But telegraph union activity was initially confined to the operators, a new group of waged workers who, preserving their popular image as white-collar laborers, refused entry to the blue-uniformed messengers. Ironically, the less likely it became for messengers to advance into such operator positions, the more the operators came to see their own interests as bound up with those of the messenger boys. Both the sheer number of messengers and the youthful eagerness of the boys to strike made the messenger boys attractive to the telegraph labor movement. But again, the timing of events and the spatiality of messenger employment were both crucial: messengers as a "technology of resistance" were only brought into a union when that union felt itself under attack, and messengers as voting union members were most crucial in New York City, where their numbers were greatest and their career opportunities most limited.[26]

Finally, chapter 10 brings the story of the messengers (and their former employers) up-to-date, describing the ultimate fates of both Western Union and ADT at the turn of the millennium—a time when the (now digital) information internetwork goes by the name "World Wide Web." One company has prospered by returning to its roots, while the other company has faded to a mere shadow of its former self. But even though the telegram may no longer be a viable information commodity, today in America and around the world, young urban bicycle couriers—not-so-direct descendants of the telegraph messengers, perhaps—have made a comeback in an all too familiar form.

The story of the telegraph messengers spans a long period of fundamental change, but there are some constant trends to rely on as guides. From 1850 to 1950, the patterns of increasing urbanization, migration, and industrialization that generations of historians have already identified clearly indicate a shift to a qualitatively different, "modern" American society. Perhaps the most interesting aspect of this messenger tale is that there was such continuity to this mostly young, mostly male, mostly "native white" low-wage labor force over such a long and transformative period of time, space, and technology.[27]

Yet this story is not just about the messengers in particular. It is also a case study in the way information labor works within information internetworks. Men and women who work to produce and provide information are implicated in four concurrent processes within their internetworked institutions: (1) the shaping of occupational identities (such as white-collar versus blue-collar); (2) the development of information products (such as instant messages or archived transmissions); (3) the evolution of technological systems (machines, algorithms, and work rules); and (4) the production of technological spaces (business offices, shop floors, and customer-service showcases) supporting the other three processes inside the internetworked institutions. In all of these ways, laborers participate in the constant, competitive "creative destruction" of the institutions of capitalism.[28]

However, it is important to remember that the lives of laborers do not begin and end at the factory door, but extend through time and space into the wider community. In this way, information workers become implicated in four more parallel processes in the wider public realm: (1) the "naturalization" of those social characteristics that are assumed to affect their job performance (such as maturity, masculinity, and femininity); (2) the shift in popular understandings of the products they produce and sell (for example, whether telegrams are seen as frightening, amusing, or empowering); (3) the changing public assessment of the technological systems within which they work (such as perceptions of the overall risks and rewards of information and communications technologies); and (4) the creative destruction of shared urban spaces of political-economic life in the wider world, *outside* of the internetworked institutions themselves (especially as internetworked communication transforms the urban infrastructure for all other production and consumption processes in the city). Together, the telegraph, post office, and telephone networks helped drive the very urbanization, migration, and industrialization processes that spawned these networks in the first place, in a complex, dialectical relationship implicating not just network designers and network users, but network laborers as well.[29]

Today, just as a century ago, human messengers hand-ferry valuable information through a recursive urban grid of streets, buildings, and offices, pushing "data packets" not with electrons but rather with their running shoes, their bicycles, and their vans. These messengers ply both the physical boundaries between virtual information networks and the metaphorical boundaries between contradictory realms in our new "informational" economy—the seen and the unseen, the indoor and the outdoor, the virtual and the physical, the child and the adult, the entrepreneur and the employee, the public and the private, the local and the global. By following an earlier generation of telegraph messengers historically through the complex and uneven spaces of yesterday's analog information inter-network, those same boundaries (and boundary workers) will be easier to perceive in the digital information internetwork we are building today.[30]

CHAPTER TWO

WESTERN UNION AND THE INTERCITY MESSENGERS

Boys, one word more! Electricity is very fast—very. It compasses the earth in a second. But it needs you at the end. It cannot do without you. You are, in many important respects, the life, the energy, the soul of the system. The postboy is a mere piece of baggage compared to you.

—From "The telegraph dispatch: A story of telegraphy in the early days," 1877[1]

These messages are flashed to the four corners of the continent, and abroad, over wires and equipment that have cost millions of dollars to build and maintain, only to be entrusted, finally to you for delivery. If an incorrect delivery is made or an unnecessary delay occurs, the speedy handling of such telegrams over great distances has accomplished nothing; the value of the service has been destroyed.

—Western Union messenger manual, circa 1910.[2]

In its first American incarnation, the telegraph was created to be an intercity system of communication. Before the telegraph, information could only be moved over long distances between cities through human, animal, or mechanical transportation. Thus the time of information transfer was limited by the physical infrastructure of trails, roads, canals, and finally railroads, all dependent on the vagaries of the weather, the seasons, and any number of other material obstacles that might arise. The application of electricity to communication created a new way to move information, changing the very definitions of speed and distance across the continent. But rather than "annihilating space with time" as its proponents claimed, new technologies of both machines and labor also "annihilated time with space," producing a new geography of information in the United States: an uneven geography of small towns, big cities, and the connections between them.

By the late 1830s, American inventors Samuel Morse and Alfred Vail had patented a simple electrical telegraph, described in the language of the times as "an insulated wire conductor uniting two stations; a galvanic battery to generate the electric fluid; an apparatus to transmit the current upon the line, called a key or manipulator; and an instrument to observe the passage of the current, called a receiver." After five years of seeking sponsors for a prototype line, they were granted $30,000 by Congress in 1843 to build a telegraph line between Baltimore and Washington, D.C. The line began operating in 1844, and by 1845, having proven its technological feasibility, it opened to the public at rate of 1¢ for every four characters sent—in other words, a rather expensive penny per word when compared to the 6¢ per page charged by the post office to travel the same distance.[3]

Congress declined at this time to purchase the telegraph, so in that same year the first private U.S. telegraph company, Magnetic Telegraph, was incorporated. Magnetic set up its first line from New York City to Philadelphia. Since no means of crossing the Hudson River with wires was available, messages had to be written out and physically relayed across, so the line was not fully automated. At first, messages were transcribed at Newark and taken to New York City by train; later, they were transcribed at Jersey City and sent across the Hudson by boat, six times a day, with young boys called "carriers" transporting the dispatches by hand to the Wall Street office. Even with these limitations, the line was a success and was soon extended to Boston. In 1846 the Boston-to-Philadelphia segment was hooked to the existing Baltimore-to-Washington line, creating the East Coast's first electric communications backbone.[4]

The Hudson River crossing was not the only reason young laborers were needed on this backbone. Magnetic personnel over the six cities in 1845 included eight "boys" out of a total of twenty workers. Company rules specified that "The Chief Operator at each station is held responsible for the employment of a sufficient number of faithful Boys promptly to deliver all Messages by day and by night. *This is an essential point and must not be omitted.* One Boy, and if necessary more should be kept in attendance at the larger offices until business closes for the night, and also on Sundays." The outdoor delivery duties that these messengers performed allowed the operators and the clerks to remain in the office, and from the start the cities receiving the most messages—New York and Philadelphia—were also the cities employing the most messengers[5] (see table 2.1).

Table 2.1. *Magnetic Telegraph operating personnel, 1845*

City	Operators	Clerks	"Boys"
Washington	1		
Baltimore	2		1
Wilmington	1		
Philadelphia	3	1	3
Jersey City	3		
New York City		1	4

Source: Magnetic Telegraph Company, Articles of association [. . .] (New York: Chatterton and Crist, 1847).
Note: New York City had no operators because messages were handwritten on Wall Street and then ferried across the Hudson to be telegraphed by operators stationed in Jersey City.

Following this model, the telegraph spread rapidly. Local telegraph companies sprouted in different areas throughout the 1850s, each hoping to be the first to bring the new technology to its region. In the early 1850s, some 450 U.S. towns were served by 1,186 telegraph offices, usually open from eight or nine in the morning to nine or ten at night; by 1857, the number of places served by the telegraph had jumped to 800, and telegraph lines extended unevenly along the East Coast all the way north to the Great Lakes (Rochester, Buffalo, Cleveland, Detroit, Chicago, and Milwaukee), all the way west to the Mississippi (St. Louis) and all the way south to the Gulf of Mexico (New Orleans). This was at a time when only about 10 to 20 percent of the U.S. population lived in cities of more than 2,500 people, and when five of those cities— New York, Philadelphia, Boston, Baltimore, and New Orleans—held about half of this small urban population. Nevertheless, new telegraph companies, following the example of the post office, believed that their quickest route to profitability was to link the largest cities first.[6]

One of these local telegraph ventures was the New York and Mississippi Valley Printing Telegraph Company, incorporated in 1851 by telegraph pioneers Hiram Sibley and Ezra Cornell (among others) to compete with the Morse telegraphs by using the printing telegraph system developed by Royal E. House. Five years later, in a buyout of several of its Morse competitors, this company was reincorporated as the Western Union Telegraph Company (hereafter simply Western Union or WU). The new company was reorganized to utilize the Morse technology, and its headquarters were moved to Rochester, New York, where they would remain for ten years. Western Union was not yet a monopoly by any means, but the Civil War soon provided an opportunity for the company to consolidate the industry. When the war began, northern telegraph lines were swamped with private, commercial, and government messages. This compelled the government to erect military telegraph lines of its own, and motivated WU to finish its first transcontinental telegraph line in 1862. But lines reaching into the South were quickly disabled. As compensation for the wartime loss of private lines, Union general Thomas T. Eckert ordered the government to hand over to private telegraph companies some fourteen thousand miles of government-built telegraph lines. Soon afterward, Eckert came to be employed by WU, the company that, not coincidentally, ended up with most of the free lines.[7]

At about this same time, financier Cornelius Vanderbilt decided to buy enough Western Union stock to control the company. Thus WU emerged from the Civil War a company with strong government ties, national reach in lines, and deep pockets of capital. The next step was consolidation: in 1866, Western Union purchased its two main rivals, American Telegraph and United States Telegraph. American Telegraph in particular had dominated the Atlantic seaboard telegraph market since 1859, when it had taken over the original Magnetic Telegraph, and after WU bought out American, the combined company moved to American's former home at 145 Broadway in New York City. Western Union would remain in this urban epicenter of communications, immigration, and trade well into the twentieth century.[8]

Western Union was now the largest U.S. telegraph company, centered in the largest U.S. city, and extending wires across the country from coast to coast. But the geography of the telegraph was more complex than this, depending not only on miles of wire but also on the time and cost characteristics of the industry's main product: the telegram. In 1852, the *Albany Evening Journal* had coined the word *telegram* in place of *telegraphic dispatch* or *telegraphic communication,* saying, "Telegraph means to write from a distance—Telegram, the writing itself, executed from a distance." Though countless new services would be offered by the telegraph companies over the next century, it was through the telegram that they were judged by their customers, their investors, and their detractors. After all, the telegraph industry was trying to do more than simply sell "information." It was marketing a new way of thinking about time and space, redefining the very terms of distance and speed. But in order to do this, the telegraph industry itself needed to imagine time and space in new ways, producing a unique physical and human geography of its own. Thus the uneven geography of the telegraph network depended not just on distance and center, but on service coverage, message price, and transmission speed—all of which were intimately related.[9]

To nineteenth-century telegraph managers, optimal service coverage was a moving target, based not only on changing concentrations of population, but on changes in the transportation infrastructure linking those concentrations. Though the 1866 triple merger was a strategy to build a national telegraph system out of three regional ones, simply accumulating offices and wire poles was not enough. The challenge of creating optimal service coverage meant keeping enough stations and lines to provide a network dense enough to attract customers, while at the same time winnowing out unprofitable or infrequently used stations—after all, the telegraph was under no state mandate to provide "universal service" like the post office was. As early as 1850, the telegraph companies began to team up with the railroads in order to provide communications nodes in railway stations while sharing rights-of-way, maintenance costs, and office labor. Typical contracts between railroad and telegraph companies divided up expenses by having the telegraph company furnish wires, instruments, and batteries, with the railroad company building, monitoring, and repairing two sets of telegraph lines—one for railroad business and one for telegraph business.[10]

Most railroad-telegraph contracts also stipulated a certain division of, and management of, telegraph labor. The railroads conveyed telegraph linemen and office workers along the tracks for free; in return, railroads were allowed to freely use the main commercial telegraph wire whenever their own telegraph wires were down. But most important for the telegraph companies, railroad depot operators were obliged to accept and process public telegrams. All payments for these messages were supposed to go to the telegraph company, but in practice, the railway operators paid themselves first out of the receipts. Over the years, different schemes for paying these subcontracted employees would be tried, but railway operators would never be counted among the regular "commercial" telegraph employees (nor would they join the same unions or engage in the same strikes).[11]

Western Union's total number of offices grew steadily after the Civil War. However, over 75 percent of these offices were joint railway offices with subcontracted operators, mostly in rural areas or small towns. In 1882, WU president Norvin Green bragged, "We pay salaries at only 2,578 of our 12,041 offices, and at 960 others we pay only a portion of the operating expense, leaving over 9,000 offices which are maintained and operated for us by railway companies in consideration of the large telegraph service performed by us for them." A decade later, in 1894, this consideration included a commission: the railway agent was allowed 10 percent of the gross telegraph receipts of the office. The situation had not changed much by 1909, when New York State had 134 commercial telegraph offices but 1,127 railway offices—with 479 of those railway offices taking in less than $2 per month in gross earnings (not even a telegram a day). Proportions were similar in other states, especially those with the greatest total number of telegraph offices, such as Illinois, Iowa, and Ohio[12] (see table 2.2).

However, relying on office totals alone can give a false impression of the extent of the telegraph system. The telegraph had always competed with the

Table 2.2. *Western Union office revenues, selected states, 1909*

State	Total Western Union offices	Percent earning over $50/month	Percent earning under $50/month
Illinois	1,433	7.7	92.3
New York	1,261	10.6	89.4
Iowa	1,100	7.2	92.8
Ohio	1,019	9.8	90.2
Pennsylvania	975	13.0	87.0
Missouri	837	7.5	92.5
Texas	775	18.7	81.3
Indiana	763	10.4	89.6
Wisconsin	646	7.7	92.3
Minnesota	610	7.0	93.0

Source: U.S. Senate, "Investigation of Western Union and postal telegraph-cable companies," Senate document 725, 60th Congress, 2nd session (Washington, D.C.: U.S. GPO, 1909), 20.

post office, and after the 1870s it also competed with the telephone. But while the post office had a branch in every town, and the telephone was designed to be installed in individual homes or businesses, the telegraph was limited in the places it could serve by the fact that it required both physical offices (housing skilled workers and dangerous batteries) and physical wire connections (expensive to build and maintain, and thus often limited to existing rights-of-way like turnpikes and rail tracks). Plus, a telegraph company would normally have more than one office in each large city, and competing telegraph companies were further forced to maintain redundant offices adjacent to each other. Thus telegraphs always reached fewer places than either the post office, and (eventually) the telephone.[13]

In those areas that did have telegraph service, what most concerned managers and customers alike was message speed. But even though the speed of light was a constant, the speed of telegraphic transmission varied widely with geography, in a fundamental conflict between urban and rural traffic. So-called through business, or traffic between the two large urban terminal points of a line, tended to monopolize telegraph resources for most of the day. This meant that "way business," or traffic between smaller stations on the line, was delayed out of proportion to the distance involved.[14]

Speed was so variable that in 1884 telegraph employees were "particularly cautioned against making any promise to customers respecting the transmission or delivery of a message." Things had improved little fifty years later, as WU's 1941 "Retail Telegraph Sales Manual" still warned that the "most frequent and natural question which confronts counter attendants is the patron's inquiry as to when the telegram will be delivered," recommending that clerks "anticipate this thought in the customer's mind by a quick and positive assurance that the telegram will be handled promptly."[15]

Somewhat paradoxically, through traffic and way traffic also differed in terms of quality. Because they had to be relayed from station to station by hand, longer-distance through-traffic telegrams were often garbled by the time they reached their destination. For example, in 1852 a telegram traveling the two thousand miles from New York to New Orleans had to be rekeyed four or five times along the way at intermediate stations. A garbled message would have to be resent, slowing overall transmission speed. Breaks and other problems also kept the lucrative through business to a minimum, even when different telegraph lines cooperated with each other—which they often didn't. All this meant that a "lightning" telegram could take hours to reach its destination.[16]

The spaces of service coverage and the speed of message transmission helped determine the geography of telegram price. The telegraph companies all agreed that the longer a message was, and the farther it traveled, the more it should cost; after all, longer and more distant messages took more time to key in, to repeat, to decode, and to (if necessary) resend. The WU triple merger had a dramatic effect on this telegram cost structure. Before the merger, each local telegraph carrier added its own rate to any through message, making the final rate difficult to determine and dependent on the actual path the message took across the country. With consolidation under WU, uniform rates were instituted. But price was still dependent on message distance, so the long-range interurban messages were favored by WU because such transmissions generated 65 to 80 percent of all revenues.[17]

Even though telegram prices, or tariffs, were theoretically constant no matter what content the telegram carried or to whom, in practice different groups of consumers received different telegram service—resulting in another differentiated geography. The first and most lucrative telegraph customers were business users: newspapers, merchants, bankers, stock brokers, and speculative investors. In 1883, telephone entrepreneur Gardiner Hubbard, pushing for government takeover of the telegraphs, estimated that 80 percent of messages were business transmissions. A year later, WU president Norvin Green agreed that only 6 percent of messages were "social."[18]

The hefty cost of sending a telegram contributed to this consumer demographic. The first Magnetic Telegraph line from New York City to Philadelphia in 1845 charged 25¢ for ten words—the same as the price of a full page of text sent all the way across the United States through the post office. But time-conscious speculators, whether "lottery men" or commodity brokers, were willing to pay for the up-to-the-hour knowledge that the telegraph promised. A customer could even ask a local telegraph operator to personally query the price of a commodity in another city—the two operators would have a telegraphic conversation, and then the first would bill the customer for the information.[19]

By the 1850s, some Wall Street brokers sent six to ten messages per day, especially between 10:00 A.M. and 5 P.M., and paid $60 to $80 per month in telegraph charges. One line reported, "The receipts from twenty leading commercial houses doing business through us average $500 each per annum." Between 1845 and 1854, commodity markets based on the telegraph, trading wheat, corn, oats, and cotton, were established in Buffalo, Chicago, Toledo, New York, St. Louis, Philadelphia, and Milwaukee. The linking of East Coast cities to the New York City financial market through both the telegraph and the new intercity express services that were started in competition with the post office, coupled with the gold rush in California and the nationwide railroad construction boom, made Wall Street a center of such speculative activity. Even with the Panic of 1857, by the start of the Civil War in 1860, Wall Street was setting prices for the whole nation.[20]

Businesses were not the only users of telegrams. As early as 1848, Morse competitor Henry O'Reilly promoted "social" telegrams, especially during holidays, "that 'absent friends' may interchange the 'compliments of the season' where they cannot enjoy the festivities of association at 'the ole homestead.'" By the early 1850s, contemporary authors listed a myriad of social uses for the telegraph: corresponding with families while traveling, sending jokes or party invitations, and announcing both births and deaths. Sometimes two distant friends would arrange a time to both be present in their respective telegraph offices, to "converse" in real time through the operators. But experiments in the early 1870s in setting up social "telegraph clubs" by wiring homes directly to each other proved too difficult for users. And for most of the public, telegram rates remained too high for regular social use. A ten–word message from New York to Philadelphia cost 30¢ in 1873, 15¢ in 1883, and 25¢ in 1908.[21]

This division between business and social uses related back to the geography of the telegraph. Very quickly, the different uses to which the telegraph was put, based on different prices for different-sized messages traveling different distances, coupled with the expectation on the part of all consumers of the greatest possible speed, forced the telegraph companies to set rules as to which messages received the fastest treatment. Most followed a scheme like that of the New York O'Reilly lines in 1852, assigning an order of precedence: government messages, "messages for the furtherance of justice in detection of criminals," death and sickness messages, important press news, and then all the rest of the normal news and mercantile messages.[22]

But in practice, lucrative through messages were still given priority over less profitable way messages. To capture such business, telegraph companies would lure large, urban commercial accounts away from competitors by offering them reduced message rates and privileged service, to the further detriment of rural customers. "Pink slips" were used to speed these special messages, which differed by region: "In lumber districts it was lumber business; in flouring-mill centers, it was flour; in fruit-growing districts, fruit; in the South, it was cotton, and in practically all places what is called 'market orders' or stock brokers' and board of trade business was rushed during the hours the stock exchanges and boards of trade were open." So even though telegraph policy was to send messages in first come, first served order, both social messages and messages from smaller places could take up to twenty-four hours to get through.[23]

By looking at the spatial and temporal rhythms of the telegraph as a whole, instead of at the electrical technologies that all supposedly operated at lightning speed, it is apparent that the intercity telegraph system had an uneven geography, with each area's particular mix of offices, messages, and customers largely dependent on the proximity of railroad stations, commodity firms, and urban centers. But just as this geography of message speed and price was overlaid on the geography of buildings and wires, so was a geography of labor superimposed upon this network as well. No matter what kind of technology was used for sending, receiving, or transmitting, the capital costs of telegraphy paled in comparison to the labor costs. For example, in 1869, Western Union president William Orton wrote that salaries accounted for 60 percent of his company's operating costs. As an 1881 article in *Harper's* noted, "Anyone can construct a line who has the money to pay for it, and the mind to do so. It is now very cheap, and when the line, battery, keys, and sounders are set up, there is very little to wear out except the operators. Acid is cheap, and the metals are cheap. The only things that really cost much are the man at the key and the man at the sounder." At WU, from the 1890s through the turn of the century, salaries for managers, clerks, and operators consistently made up half of the operating expenses—more with messenger wages tacked on. These workers supported WU's main business of message transport, which made up the majority of the company's revenue through this period[24] (see figure 2.1).

Operators were arguably the most important employees. At first, two operators were needed to receive a single Morse transmission. The message itself was clicked out in Morse code onto a long paper strip by a receiving device. One operator would read the Morse and translate it out loud, and the other would hand-copy the English text of the message onto a message blank as it was spoken. With the development of "sounders," which audibly clicked out incoming Morse transmissions, most offices stopped using paper recording receivers by 1852. Now only one employee was needed in an office to receive messages, listening to Morse coming through the sounder and transcribing it directly into English on a message blank. But still, sender and receiver had to both be on the line simultaneously, and both had to have comparable skill and speed.[25]

Figure 2.1   *Western Union finances by category, 1893–1908*

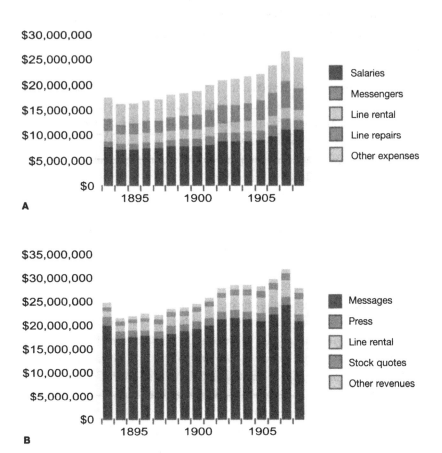

A: Expenses. B: Revenues.
Source: Western Union Telegraph Company, statistical notebooks prepared for Robert C. Clowry, 1899–1904, Western Union Archive, 1993 addendum, series G, box 81, folder 4.

For long-distance messages, where the tariff was greatest, labor was even more important, as messages had to be transcribed and then repeated by hand without error several times along the way. But labor continued to be an issue even after the first "automatic repeaters"—battery-powered devices which acted like amplifiers to simultaneously receive a message on one wire while boosting it along its way on a different wire—were introduced in 1869. As late as 1894, one telegraph manual cautioned, "it requires the service of a first-class operator at each repeater to keep it in adjustment. In long circuits, repeaters are employed about every 500 miles, at each of which new battery force is supplied, so that on a circuit of one thousand miles through repeaters, three operators are employed." Such repeaters reduced operator effort, but the operator was still present.[26]

Operators weren't the only telegraph employees. Labor forces varied depending on the size of the local office, and might include everything from low-paid clerks, "check-boys," and route aides, to high-paid traffic managers, circuit managers, chief operators, and divisional superintendents. In 1866, at the time of the big WU merger, the *Telegrapher* counted the telegraph offices and employees in selected cities, and found that the number of messengers generally equalled the number of managers and operators in the most important urban markets. Their count omitted "battery men" and janitors who worked part-time, but the explanation noted, "The force of a single main office, having six wires, in a city like New-York, would be, for day service, 1 manager, 3 receiving clerks, 8 operators, or 12 if service is continuous, 2 copying clerks, 1 office boy, 1 delivery manager, 3 bookkeepers, 16 messengers, 1 battery-man, [and] 1 repairer." The official Western Union count of messengers versus operators across the nation from 1867 to 1879 showed a significantly smaller percentage of messengers, attributable both to the smaller percentages of messengers in rural areas versus big cities, and to the subcontracting of messengers from other companies in those big cities (discussed in chapter 3). But even in this conservative estimate, messengers made up 15 to 20 percent of Western Union's nationwide workforce[27] (see table 2.3 and figure 2.2).

After the 1866 triple merger, WU president Orton tried to bring a certain technological coherence to the labor force of his new company. According to one contemporary, Orton's ideal was that "in order to [achieve] unity of administration, distinct and clearly-defined ideas of duty should be made to permeate the entire working force, so as to make conflict impossible, and work quick, certain, harmonious." WU even advertised in 1869 that "the operation of our system over the vast territory covered by our lines is fast assuming the certainty and uniformity of mechanism." But such a harmonious state was far from reality; there were vast disparities in wages and skills between offices in different areas. At many rural railway offices, operators could not make a living from telegraph duties alone, and had less time to devote to telegraph traffic, less time to hone their telegraph skills, and less money to pay for assistants like messengers who could increase the physical range of the office. A station agent from Utica, Pennsylvania noted with pride in 1870 that he filled seventeen jobs: "railroad agent, freight agent, ticket agent, station baggage man, clerk for railway, porter for railway, agent for United States Express, money clerk and porter; manager for W.U. Tel. Co., day operator, night operator, receiving clerk, forwarding clerk, error clerk, message boy and porter."[28]

This labor difference had a gendered aspect as well. Western Union relied more on women in its smallest offices, such as rural or hotel offices, where the

Table 2.3. *Western Union division of labor, eastern cities, 1866*

City	Offices	Managers	Operators	Clerks	Messengers
New York	74	5	202	109	168
Philadelphia	35	4	78	43	86
Baltimore	19	4	42	20	48
Washington	16	4	43	30	38
Boston	24	4	58	34	60
Chicago	22	2	36	12	36
Cincinnati	21	3	43	10	37
TOTAL	211	26	502	258	473

Source: *Telegrapher* (September 1, 1866)

Figure 2.2. *Western Union national division of labor, 1867–1879*

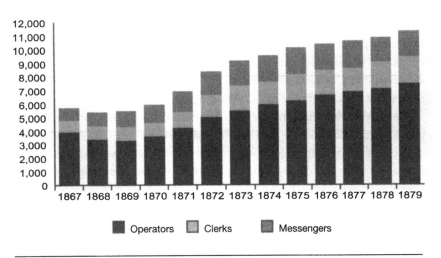

Source: WU, record books of George Prescott (1867–1912), New York Public Library manuscript collection.
Note: Does not include messengers subcontracted from district telegraph companies (a practice beginning in 1872).

low salary of the female operator (about half of what a man might make) could be paid entirely out of the office's receipts. Thus offices run by women were less likely to have messengers. Women were also assumed to be less skilled "on the key" than men, so were often provided with automatic equipment. But relying on automatic equipment before the turn of the century was much slower than employing a skilled Morse operator. Once again, either the speed of commercial messages suffered, or these workers would be relegated to lines and offices where speed was thought not to matter.[29]

What about the messengers? As described earlier, young boys were a part of this labor "mechanism" from the start; in fact, the number of messengers that an office was able to employ was a good indication of that office's "rank" within the telegraph network. In 1848, the operator of one small office on a branch line through Michigan was told that he could not have a messenger because only four messages were sent per day, and only five or six were received (the office took in only $20 in message receipts and $28 in press reports per week). As the telegraph spread and volume grew, more and more offices needed the boys. But there would always be a geographical disparity, with large urban offices needing more boys than small rural ones.[30]

The spatial allocation of messengers was dependent on the temporal demands placed on each office. In offices with regular traffic, especially through business, which had to be passed on down the line, an operator was obliged to stay in the office and on the key at all times. This meant that someone else had to attend to the rest of the tasks of running the office—not only delivering telegrams to customers, but obtaining supplies, collecting the bills, and running miscellaneous errands. Messenger boys filled all of these tasks, within a spatial and temporal division of labor constructed so that the higher-paid, skilled operator never had to leave the key.

Delivering telegrams was of course the primary task of the messenger. Such "final delivery," as it was called, involved getting to the customer's location, finding the customer, handing over the message, and obtaining a signed receipt. The messenger was expected to solicit the customer for an answer as well, meaning that even though customers had to go to telegraph offices to send telegrams, they could respond to telegrams from the comfort of their own homes or offices. Before 1872, there was no way to summon a messenger on demand to pick up a telegram, transport a message, or perform an odd task; the best that an important customer could do was to demand that a messenger make daily or hourly visits (much like a post office carrier) in order to transfer telegrams in batch.[31]

Early telegraph rulebooks show that telegram delivery was not a simple matter, and as late as 1922, educators warned that messenger duties were "not so simple as they seem," with "a right and a wrong way of doing them." Messengers first had to know how to find their way around their city. When multiple messages were to be delivered at once, a "route clerk" was supposed to figure out the most efficient route and arrange the messages in that order—messengers were admonished to "Follow the route indicated by the route clerk and do not waste time in making the deliveries." But route clerks could be mistaken, if they bothered to do their jobs at all, and messengers were on their own once they left the building. Often messengers would have to inquire not only in building management offices, but also at the local post office and later at the local telephone exchange for use of their street directories. The job demanded both spatial skills and problem-solving skills.[32]

Finding the address was only the first step. At first, telegrams could only be delivered to the addressee, a practice that became known as "HTA," for "handed to addressee." After 1870, a messenger was allowed to leave a telegram with a household servant, but then a duplicate telegram was to be delivered to the recipient's business office the next day. Even if the recipient was available, the messenger still had to solicit a response, which meant providing a pencil, telegram blank, tariff sheet, and correct change to the customer—and avoiding the temptation to overcharge.[33]

Some level of trust was necessary. Telegrams were picked up and carried to the district telegraph office by messengers "free of charge," but unless the customer was a billed subscriber, the customer still had to pay the messenger for the final cost of transmitting the telegram along the wire. The messenger had to collect the correct amount of money from the customer, based on distance, number of words, and type of message. Even if the message was not a telegram, but merely a handwritten slip of paper traveling across town, the calculation involved a complicated cross-referencing with the tariff directory for that city. And if multiple messages were traveling in the same direction, one had to calculate the cost to the farthest destination and add 5¢ for each stop along the way. This was a tedious procedure, and even though the back cover of one directory was emblazoned with the warning, "NEVER PAY MONEY TO MESSENGERS WITHOUT FIRST CONSULTING THE TARIFF," in practice, customers would often trust the messengers to compute the correct charge—and managers would lament that the messengers "at the best turn into the company but 50 per cent. of the profits of their service."[34]

Finally, every message that was not delivered on the first try had to be dealt with. Messengers had to report back with any undelivered telegrams, and were not excused from delivering business telegrams once the destination office closed for the day. Customers could file their home addresses with the telegraph office, and messengers were expected to attempt delivery after-hours. Each failure to deliver a telegram was recorded by the delivery clerk. And when telegrams could not be delivered, notice was to be sent to the addressee through the post office.[35] (This is explored in more detail in chapter 7.)

Telegram delivery took up a good portion of the messenger's day, but all messengers spent some time just hanging around the telegraph office. The companies put these "idle" messengers to good use, the start of what would grow to become a sophisticated "load-leveling" strategy later. The notion of load-leveling relates to the economic rationale of building and operating technological systems in order to accumulate profit: letting machinery sit idle during certain parts of the day, week, or year is inefficient, since the owner pays for that machinery whether it is in use or not. Thus anything that can move peak effort into the trough times, leveling the load, so to speak, will not only utilize existing machinery more efficiently, but will delay the need to purchase further capacity. The same holds true for messenger labor: the more tasks messengers could perform during the slow periods of the day, the more efficiently each messenger's labor would be utilized. For example, before the advent of power grids and fuel-powered dynamos, telegraphic current would be supplied by chemical batteries in each local office, batteries that had to be maintained regularly. This dangerous job initially fell to messenger boys, though it was later handed over to "battery men." And as late as the 1870s, messenger boys did line repair work as well. Their mobility made them the cheapest labor to spare in what could be a long walk along the telegraph line.[36]

What were messengers paid for all these services? Their wages in the early days of telegraphy were low, averaging between $3 and $6 per week all the way through the turn of the century. This was for a job that demanded long hours of attention—some worked seven days, sunrise to 10:00 P.M., except for morning and afternoon church hours on Sunday. Future industrialist Andrew Carnegie, perhaps the most famous messenger boy of them all, worked until 11:00 P.M. on alternating days, and until 6:00 P.M. during the rest, when he was a messenger in 1850. The only holiday came for two weeks during the summer. He was paid his wages monthly, earning $11.25 per month. Low messenger wages were sometimes supplemented by bonuses. For Carnegie, messages delivered beyond a certain distance (most likely, one mile) were termed "dime messages" because the boy earned a dime extra for each one. But bonuses or not, from 1867 to 1876, the average wage for Western Union messengers changed little, hovering between $3 and $4 per week at a time when operators earned two to three times that amount[37] (see figure 2.3).

These wages seem small by today's standards, but were they poor wages from the point of view of the messengers at the time? One indication that they were is the fact that messengers regularly devised schemes to eke a few more cents out of the company, the customer, or both. As early as 1845, on the very first commercial Magnetic Telegraph line that reached New York City, a manager reported, "These young rascals soon started a private banking business by having a wood-cut made with '25 cents due' thereon, which they stamped on messages and collected, and which they carried on for some time before they were discovered." Such scams continued to surface, and an 1868 writer advised thwarting them with special envelopes for telegrams marked "This message is prepaid."[38]

In any case, the messenger wage was a tiny part of the telegram's cost. In 1886, a former telegraph manager estimated the average cost of handling a message at 25¢, when the average charge per message was 32¢. Thus the tele-

Figure 2.3.   *WU weekly wages for messengers and operators, 1867–1876*

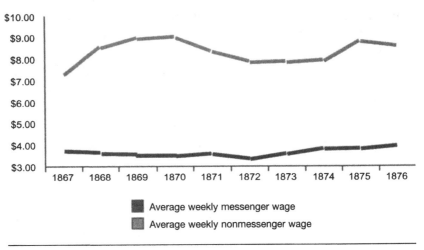

Source: WU Prescott records (1867–1876).

graph companies made an average 7¢ profit per message, at a time when messenger boys were paid an average piece rate of 2¢ per message. So even though total wages made up 60 percent or more of telegram costs, messenger wages alone made up less than 10 percent.[39]

A clue to the importance of messenger boys to Western Union in the nineteenth century can be found in the surviving record books of the company's Harrisburg, Pennsylvania office, listing the wages paid to each employee from 1865 to 1902. In terms of office personnel, there was only one manager and one janitor at a time, but other employee numbers showed different patterns before and after 1880. Before 1880, there might be one or two clerks; after 1880, there were always three or four. The number of operators varied from one to nine before 1880, but settled down to six or seven after 1880. And before 1880, there might be three or four messengers; but after 1880, the number of messengers grew steadily, reaching a high of nineteen messengers in 1901 before the messenger job was subcontracted out in 1902[40] (see figure 2.4).

The wage data for this office shows a clear hierarchy. The manager made the highest wage, at nearly $100 per month. Operators came next, making between $40 and $60 per month. Before 1880, clerks made between $20 and $40 per month, but after 1880, that average fell to only $20 per month. Finally came the messengers, making between $10 and $14 per month. All in all, messengers made up 20 to 50 percent of office workers, but only made up 5 to 33 percent of the office payroll. And the data from this office also sheds some light on what the messenger job was like. From 1882 to 1902, figures for both indi-

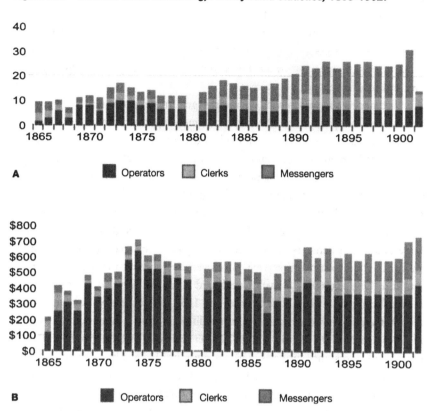

Figure 2.4. *Western Union Harrisburg, Pennsylvania statistics, 1865–1902.*

A

Operators Clerks Messengers

B

Operators Clerks Messengers

A: Division of labor. B: Yearly salaries.
Source: WU Harrisburg, Pennsylvania office ledgers (1864–1902). WUA, series 3, box 11, folders 2:3.
Note: No data for 1880.

vidual messenger wages and the number of messages handled by those messengers show that the messengers consistently earned under 2¢ per message. The boys each handled an average of twenty to thirty-five messages per day over a fifty–hour week.[41]

The construction of messenger boys at these low wage rates was not inevitable. There were significant economic constraints operating on the early telegraph companies that pushed them to seek the lowest labor bill, it is true; but in addition, there was a conceptual value being placed on the worth of a telegraphic message—and its messenger. In 1853 at the National Telegraph Convention in Washington, D.C., Magnetic Telegraph president William Swain said disdainfully that the U.S. telegraph industry had become "a very cheap business" because of the use of low-waged boys, saying, "In England, it is otherwise. Their rates may be too high, but there you find educated and intelligent men to wait on you. They charge you a good price. They deliver your message, not by a poor boy picked up from the streets to deliver messages at a small weekly pittance, but by a man—a porter—who charges his porterage, cab hire and insurance, and takes a receipt."[42]

Pittance or not, just as with operator wages, messenger wages varied greatly over space. The question of whether to pay messengers a piece wage, an hourly wage, or a salary was left up to individual offices. A St. Louis office manager argued that "regular salaries secure the most complete, and satisfactory, and vigorous service, and are, at the same time, more influential in improving the personal morale of the messenger force." Telegraph journal editor J. D. Reid agreed that a straight salary "has the advantage of placing a boy on the pay-roll of his office, and thus, to that degree, identifying him with it, and affording him the hope of advancement," while a piece rate "separates and reduces the system to an outside and contingent service." But even Reid believed that the piece rate was still "best adapted" for eastern seaboard cities like New York, where it was in fact the norm. In New York City around 1852, messengers worked on a piece rate that varied with the distance of the delivery: within a mile radius, 2¢ to 3¢ per message; at night or over a mile, 12¢ per message (including transportation costs); and to distant boroughs, 25¢ per message. Thus, a boy earned between $2 and $4 per week, a figure that would persist up until the turn of the century.[43]

In the early days of telegraphy, though, a career link between the operators and the messengers served to legitimize such low wages. For many telegraph boys on the first lines in the 1850s and 1860s, there was a real possibility for advancement and mentoring—almost an apprentice relationship. In 1848, the New York, Albany, and Buffalo telegraph line paid messengers $4 per month plus board, rather than a piece wage, in order to keep them constantly on duty seven days, sunrise to 10:00 P.M. (except for morning and afternoon church hours on Sunday). Such arrangements, while harsh, served to train the next generation of skilled telegraph operators.[44]

Apprenticing resulted in an early tradition of small, one-person rural telegraph offices being staffed by very young children and teens, such as the nine-year-old Western Union messenger boy who became an office manager in Hollidaysburg, Pennsylvania, at age ten in the 1860s. Operators were proud of such children, writing glowing letters to the telegraph journals about young "first-class operators" aged ten to twelve through the 1870s: a 12–year-old WU operator in Carbondale, Illinois; a ten-year-old messenger boy from Fremont, Nevada, who became an operator at age eleven; and a Wayne County, New York operator only twelve years old.[45]

Child telegraph operators emerged for several reasons. The Civil War provided both an opportunity and a necessity for young boys to work the wires. But after the war, young operators usually started out as messengers, and were mentored by an adult in the office—often the child's parent or other relative. In small rural railway offices, where the telegraph companies rarely paid salaries (instead offering operators a percentage of the receipts), the use of extra family labor had a dual purpose. Rural adult operators were both teaching their children a trade and making use of their children as low-cost assistants.[46]

There were limits to the pride in the youth of these operators, however. As early as 1873, the telegraph journals began to question the practice of having children under ten work at such jobs, to the exclusion of school: "Let us spare the infants," wrote one editor. Others began to fear that child telegraphers would be used to substitute for adults who weren't performing their own duties—for example, operators leaving their offices and "experimenting with Dutch Bill's beer . . . in a saloon around the corner."[47]

Yet glowing reports of child operators continued in the 1880s, describing among others a nine-year-old girl telegraph operator in Brown County, Texas, a twelve-year-old boy operator from Chester, South Carolina, and an eight-year-old boy operator from Strasburg, Virginia. Ironically, such tales undermined the idea that telegraphy was a demanding, adult skill worth the high wages that operators desired. But operators could rationalize the feats of children with a spatial and temporal argument: working the slow lines of a rural way station might be a child's job, but working the high-traffic urban trunk lines remained adult work.[48]

How difficult was it for a young messenger to learn Morse code? One telegraph manual from 1860 described Morse as requiring not only practice but superior timing, meaning young persons were thought to be able to pick it up rather easily. But another manual from 1894 warned that "constant and persistent practice is *absolutely necessary* to success." To avoid being labeled a "plug," or a poor operator, one needed an average speed of about twenty-five words per minute (though rarely could any operator send over forty words per minute). A decent job could be had at about thirty words per minute.[49]

In an office with two or three messengers, a manager might provide his boys with unused Morse keys wired together, encouraging them to practice during off-hours. Or lone messengers in way-station offices might practice with their counterparts in nearby towns, if they were given permission to use the wires at night when traffic was light. When parents employed their own children, Morse training was almost certain. But the time needed for such training could only be spared in a low-traffic office. Urban areas lacked such opportunities, and operators in larger offices regarded questions from messengers as bothersome. In 1871, the *Telegrapher* sarcastically suggested to messengers, "To gain the good will of an operator hang around him at all times, follow him from table to table, and if not immediately employed sit in his lap." Thus messenger apprenticing varied with both office size and office location.[50]

So how were operators produced in large cities? After the Civil War, private schools emerged to train young men in this exciting new field. Such schools were dubbed "ham factories" by skilled operators for their overenthusiastic promises. An 1868 circular from Porter's Telegraph College, in Chicago, claimed that graduates would have lucrative salaries and unlimited career and locational mobility:

> From three to four months spent in the telegraph college will enable young men of ability to manage a telegraph office. *The telegrapher may select his place of business and residence in accordance with his fancy, either in the great metropolis, the quiet village, the fashionable watering-place, or the isolated cabin on the plains.* . . . The study and practice of telegraphy is not tedious, but, on the contrary, to those having leisure hours, it would serve as a pastime rather than as a task.

Such glowing assessments of telegraphy continued for decades—in 1884, four months of training, including board and washing, cost $102 at the Pennsylvania and New Jersey Telegraph Instruction Company. The promises of the "plug schools" helped to attract messengers to the industry as well, hoping to learn the "pastime" of telegraphy on the job while earning only $8 to $16 per month.[51]

A tension arose between operators and messengers as telegraphy grew: on the one hand, many operators themselves had started out as messengers and ended up managing messengers, and so looked upon messengers with affection; but on the other hand, operators knew that messengers learning telegraphy might soon be hired on the key at lower wages, undermining existing operator jobs. When the choice was between accepting recruits from company-sponsored "plug schools" or apprenticing the messengers themselves and having some control over the craft, which did the operators prefer?

The answer depended on the operators' (real or perceived) chances to advance in telegraphy. The first generation of telegraph operators had a better chance at moving up than later ones did; according to one telegraph historian, "the 1870s, 1880s, and 1890s seemed the dark age of the craft. What had once looked like a boundless horizon now appeared to be a dead end as opportunities for ambitious young men within the industry shrank." The hopes of advancing from operator into some sort of inventing or engineering position had also dwindled by the 1880s, when the college-educated upper middle class began to choose this field. Thus the Brotherhood of Telegraphers, an early telegraph union of the 1880s, opposed the telegraph schools because "hams" undercut wages and lacked skill. The union demanded "the supplying of operators from the ranks of deserving clerks and office boys—the only students who are qualified to succeed us."[52]

Figure 2.5. *Telegraph messengers by state, 1910–1950*

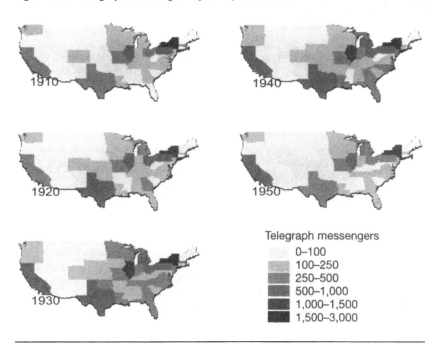

Telegraph messengers
- 0–100
- 100–250
- 250–500
- 500–1,000
- 1,000–1,500
- 1,500–3,000

Source: U.S. Census (1910–1950).

As the intertwined fortunes of messengers and operators across the nation illustrate, at the intercity scale, an uneven labor geography existed early on in the telegraph system, superimposed on the uneven geography of offices, wires, and way stations. Small towns were served by fewer offices than big cities, the speed of small-town traffic suffered at the expense of big-city traffic, and those small-town offices employed fewer and less-skilled operators than their big-city counterparts; thus small-town operators were less likely to want to split their meager earnings with messengers. Yet these same small-town offices were the likeliest sites of training for messengers wishing to graduate to jobs as operators.[53]

This uneven messenger geography persisted into the first half of the twentieth century, as illustrated by data on telegraph messengers from the U.S. Bureau of the Census. Plotting raw numbers of telegraph messengers per state clearly shows a bias toward the eastern, urbanized states, with New York, Pennsylvania, Ohio, and Illinois leading the nation in the industrialized Northeast, and Texas and California as the only real concentrations of messengers in the South and West (see figure 2.5). Similarly, plotting telegraph messengers per one hundred square miles shows that in terms of area, the boys were concentrated in the East as well, following urban concentrations of population in the nation (see figure 2.6). Plotting telegraph messengers per ten thousand

Figure 2.6. *Messengers per one hundred square miles by state, 1910–1950*

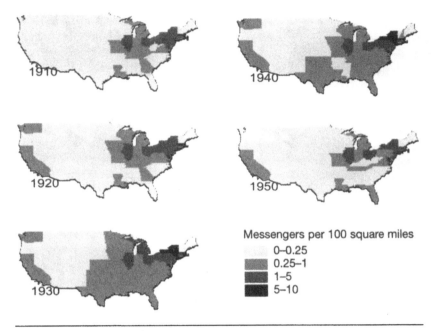

Messengers per 100 square miles
- 0–0.25
- 0.25–1
- 1–5
- 5–10

Source: U.S. Census (1910–1950).

Figure 2.7. *Messengers per ten thousand inhabitants by state, 1910–1950*

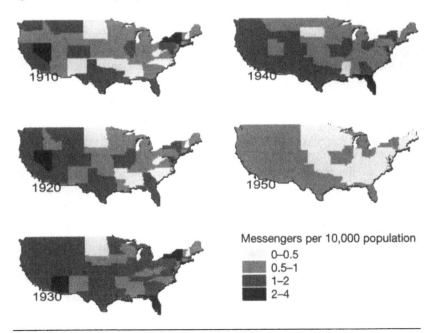

Messengers per 10,000 population
- 0–0.5
- 0.5–1
- 1–2
- 2–4

Source: U.S. Census (1910–1950).

inhabitants, on the other hand, shows that when adjusted for population density, areas of the far West actually had more messengers per capita than some eastern areas, a necessary cost of moving messages great distances from rural telegraph offices to small, dispersed populations (see figure 2.7).

Together, these statistical maps of telegraph messenger labor illustrate something that physical-plant maps of telegraph offices or telegraph wires fail to reveal: the uneven geography of the national-scale telegraph network, which was established through merger and expansion in the nineteenth century, persisted well into the decades of Western Union monopoly during the twentieth century. As the next chapter shall demonstrate, at the local scale as well, where individual communities hosted a myriad of intracity telegraph systems, the uneven geography of the telegraph was both rooted in and reflected by messenger labor.

CHAPTER THREE

AMERICAN DISTRICT TELEGRAPH AND THE INTRACITY MESSENGERS

The Messengers of this Company are chosen for their honesty, intelligence and general adaptibility to our business. The general behavior and standing of each Messenger is a matter of constant supervision by the officers of the Company, thus securing to the public the services of the finest, most intelligent, and honest corps of messengers to be found in any city of the world.

—American District Telegraph Company, 1875[1]

Soon after the regional telegraph companies began wiring the nation between cities, local telegraph companies emerged to try to wire businesses and homes together within cities. As early as 1845, a short intracity telegraph line was erected in New York City by future Western Union pioneer Ezra Cornell, attached not to poles but to buildings; however, the public was not very interested. A similar fate befell Henry Bentley ten years later. In trying to put together his own intracity telegraph system in New York City, he came up with a system of installing boxes in public places, such as drug stores, where outgoing messages could be deposited. Customers were to purchase small postagelike stamps as payment, and messenger boys would pick up the messages in batches and bring them to the telegraph office. But underpayment in stamps and illegible script caused the project to fail. Neither an all-electric nor an all-physical system worked within New York City.[2]

At the same time as such publicly accessible projects were failing, private firms were installing their own in-house intracity telegraphs. As early as 1850, large businesses began setting up private lines using Morse equipment and employing skilled operators. The cost of such systems fell with the development of automatic telegraph equipment that could be used by an unskilled operator, such as the "dial telegraphs" of the 1860s. By the 1870s, new tranceivers had appeared, equipped with pianolike keyboards and tape printers, requiring only a sending operator, not a receiving one. Such machines cost from $125 to $250; however, not only were these devices slower than Morse systems, but they usually required batteries, which was both an additional labor hassle and a hazard that could result in damage to carpets and furniture when the acid was replaced. A market still existed for a low-cost and low-maintenance intracity telegraph service that could be easily used by private and public patrons alike.[3]

In 1872, the American District Telegraph Company in New York City became the first successful intracity telegraph service, operating almost as the converse of the intercity telegraph service of Western Union. The key to ADT's system was the "call-box," invented by Edward Calahan in 1871 (see figure 3.1). ADT installed these small mechanical units in subscribers' homes; each unit was then linked by wire to a nearby district office, with about one hundred boxes sharing a single circuit. Call-boxes were run by weights or springs, not batteries, so subscribers didn't have to manipulate dangerous chemicals; the district station stored the main batteries that kept current on the line. To send a message, a customer set the call-box dial to "messenger" (the default setting) and pulled a lever. The call-box would click out the customer's unique ID number onto the line, and an instrument similar to a private-line automatic telegraph printer would print out that number on paper tape down at the main office. Then the call-box attendant would look up the number on a set of tiny wooden drawers, and retrieve the customer's name and address from a slip inside—along with the complete history of the customer's subscription payments and message deliveries.[4]

Calahan originally developed the call-box system to be used cooperatively by households in case of burglaries, without an intervening district office: "each residence (there were at least 50) should be connected in one metallic circuit. A bell in each house, on the fire-alarm order, would wake up the whole village, and the signals would indicate whose house had been entered, or who was in distress. Under such circumstances, neighbors would come personally or send their coachmen or servants, and render whatever assistance might be necessary." But under ADT the system changed from a public alert to a private service, used for general communication as well as for safety. Upon receiving a call at a district office, a young messenger would be sent to the customer's home or business, ready to transport a written message to its destination anywhere in the city. In this way, unlike the intercity telegraph, the intracity system put only control information on the wires and left the actual message in print to be transported by humans. Call-box service started at $2.50 per month—a lucrative business, since each call-box only cost about $6 to produce.[5]

District telegraph service spread quickly to other cities. ADT in New York

Figure 3.1. *ADT San Francisco call-box, 1883*

Source: ADT San Francisco, *Directory* (1883).

controlled the original Calahan patents on district call-box systems, and licensed these patents to similar independent systems in other cities—most calling themselves "American District Telegraph" as well. Competitors soon emerged, controlling other patents, such as the Domestic Telegraph Company, a district service created in 1874 and relying on Thomas Edison's patents, not Calahan's. But in all cases, messenger boys were key to the system's success.[6]

With two scales of telegraph systems, one within cities and one between cities, it soon became obvious to the builders of both that there were advantages to linking the two of them together. In March 1873, ADT vice president E. B. Grant gave WU president William Orton a demonstration of the call-box and messenger system in New York City. A signal sent from the "farthest point" of the test district reportedly resulted in a messenger arriving three and a half minutes later. After this meeting, WU itself became an ADT client, with four call-boxes rented for the WU executive suites. Six months later, ADT fired its own telegraph operators and started making space for WU operators instead. Clearly an agreement was in the works.[7]

In November 1873, efforts began at WU to determine "what may be saved by combination with the American District Telegraph Company." By April 1874, an agreement was struck "to increase the efficiency of the service of the two Companies and not to create a profit for the one at the expense of the other." In any city where both ADT and WU had offices, ADT would place a call-box in each WU delivery office, supplying messengers "when required."

ADT would then deliver WU telegrams for 5¢ each, returning answers to WU for 5¢ more. This was apparently the fee to ADT, not the charge to the customer, to whom telegram delivery was "free" in New York City. (Of course, the messengers earned substantially less than 5¢ per message.)[8]

That first contract between WU and ADT in New York City set a pattern for interrelationship between the two companies in all other cities to follow. But although the key motivation for the agreement was the economics of telegram delivery (with WU wanting to control delivery costs by outsourcing them at a fixed rate, and ADT wanting to capture a lucrative and steady source of revenue from the delivery business), a detailed reading of the contract shows that there were three other important (and familiar) reasons for the companies to join forces.[9]

First was *increased service coverage*. On the one hand, WU turned every ADT office into a WU office, since all ADT offices were to accept incoming WU telegrams from customers free of charge, forwarding these telegrams by messenger to the nearest WU office. But ADT increased its spatial extent as well, since WU and ADT would jointly occupy six new offices around the city, with rents and expenses shared.

Second was *increased message speed*. In order to speed delivery of intercity telegrams, ADT agreed to increase its number of delivery districts while decreasing the extent of each, meaning it would open more offices and, quite likely, employ more messengers. But WU also helped to speed ADT intracity message delivery, by arranging "joint offices to be so connected by wires that intercommunication between them shall be practicable without repetition through the Western Union main office." This meant that in each joint ADT/WU office, ADT intracity messages could be sped part of their distance along the wire and would not compete with (or be slowed down by) the intercity lines leading from the joint district office to the central WU office.

Finally, the comapnies could *reduce labor costs*. WU realized it could reduce the number of its offices employing "first class operators" by staffing some offices with receiving clerks only, and having messengers run incoming telegrams from these offices to the operator offices. This meant that low-cost messenger labor would be used to save on expensive operator labor. Similarly, ADT would keep one messenger in each of sixteen other specific WU offices around New York City. This meant that each messenger could be used for both WU and ADT business—both intercity telegrams and intracity print messages.

Thus at the end of its second year of business, ADT New York City had negotiated with WU to share responsibility for 28 offices, and any of ADT's 2,500 subscribers could call for a messenger who would forward a scribbled note to WU as a telegram. Within New York City, ADT and WU claimed that 80 to 90 percent of the population would be "within three minutes of such an

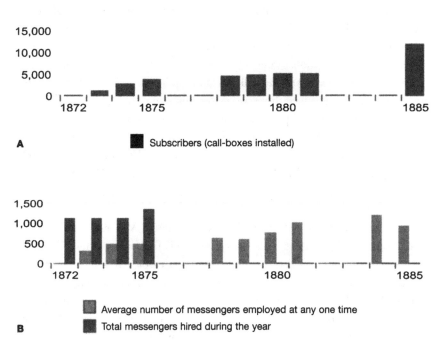

Figure 3.2. *ADT New York City statistics, 1872–1885*

A

■ Subscribers (call-boxes installed)

B

■ Average number of messengers employed at any one time
■ Total messengers hired during the year

A: Call-box subscribers. B: Messenger employment.
Sources: Reid (1886); *Telegrapher* (1873–75); *NYT* (1875–80); *JoT* (1879–80); Johnston (1880); Oslin (1992); *Operator* (1881).
Note: Blank values indicate missing data.

office." By 1884, ADT had 26 different district offices in New York City, and delivered telegrams for free to the closest of 147 Western Union offices in the same area (claiming an average delivery time of seven minutes). And as their call-box network grew along with their messenger force, so did their profits— by the early 1880s, ADT New York City had some 12,000 subscribers, employed about 1,000 messengers, and cleared about $100,000 in profit each year (on receipts of about $500,000)[10] (see figure 3.2).

With the innovation of the district telegraphs, a new category of customer was created: the call-box subscriber. For the first time, a telegraph consumer committed to paying a monthly fee for the ability to summon a messenger, whether that consumer planned to send intercity telegrams, intracity physical messages, or both. A call-box subscription only ensured access to the messenger—customers still had to pay charges for each message. At first, the district companies sought to optimize (and not to maximize) their subscriber base within each city, just as the national companies sought to optimize (but not maximize) the towns where their offices were located. In 1878, ADT Philadelphia touted their "careful weeding out of boxes in the hands of non-paying, doubtful or unprofitable subscribers"—they were in no way providing universal access to their system. But as contracts multiplied with national companies like Western Union, gradually the district companies began to maximize access to their service, installing call-boxes in public places like hotels, restaurants, and drug stores, since those were sites where many telegrams originated. As telephone exchanges expanded starting in the 1880s, more and more call-box subscribers switched over to phone subscriptions, allowing them to summon messengers just as easily. Even so, by 1927 WU still had 258,000 call-boxes nationwide, with 49,000 of them (20 percent) in New York City. About 80 percent of all New York City telegrams were picked up by messengers in this way.[11]

Enamored with the call-box and messenger system, Western Union set up subcontracting relationships with independent ADTs in many cities. By 1884, WU president Norvin Green reported while WU employed only about 2,900 messengers of its own nationwide, it subcontracted with district telegraphs for another 5,000 boys—meaning a total of about 8,000 messengers made up nearly 38 percent of all WU employees. In 1902, WU even organized all the various local ADTs around the nation under a new holding company, valued at $10 million and incorporated in New Jersey. The new national American District Telegraph Company traded stock for stock with all the local ADTs in order to centralize ownership. Western Union had now solidified its control over the intracity telegraphs: WU president Robert Clowry became president of the new ADT, "appointing all Western Union managers to be managers of the new company in their respective territories." In 1910, Western Union officially absorbed these local ADT franchises.[12]

Western Union had succeeded in rationalizing the business function of "final delivery," subcontracting delivery labor at a low, fixed cost per message and externalizing problematic issues of surveillance, turnover, and career advancement. But the call-box system meant that messengers were now more than simply technologies of final delivery. ADT messengers redefined the range of duties that boys could perform for the telegraph industry. The ability to summon messengers on demand suddenly made them useful for the general delivery of packages of all sorts. In the mid-1880s, said one WU official, "Carrying letters, coupons and valuables is their chief employment in lower New York." This was telegraphic "information" that never made it into the electrical system. By 1917 in Philadelphia, for example, the three hundred messenger boys handled seven hundred "non-telegraph messenger services" per day. Deliveries of "money, legal documents, machine parts, movie films, and other articles needed in a hurry" persisted for over fifty years. Usually paid a piece rate for this task, messengers were cheap enough and reliable enough that local merchants would hire them as needed, instead of keeping their own delivery or errand boys on hand. San Francisco messengers even delivered coal.[13]

These activities had new effects on the spatial extent of the telegraph. Though telegrams would be picked up and delivered only within a certain radius of a messenger's own branch office, package pickup and delivery could send a messenger far and wide. In 1908, a Chicago reporter commented that a messenger's delivery tasks "may lead his footsteps either to the heavily curtained drawing-rooms of disorderly houses in the Red Light district or to the wet planks of the wharves on the Calumet River twelve miles away, where he will curl up under the stars and sleep till the delayed boat arrives from Duluth." Western Union vice president W. S. Fowler agreed in 1924: "His eight hours may be spent in routine deliveries to the neighborhood; or he may get an out-of-town assignment and be sent to some distant city to deliver a package or document." The low hourly rate paid to messengers made such distant overnight duties possible.[14]

Messengers now competed not only with other outdoor delivery boys, but with indoor office boys as well. District telegraph companies advertised "*Trained, Uniformed and Reliable Messengers for any Service*" as early as 1884, since the call-boxes placed in businesses provided a convenient way to instantly summon boys for hire by the hour. Again, such services worked to the advantage of Western Union, because each time an ADT boy was hired by a business for an odd job, that business was more likely to later use that ADT boy (an official agent of Western Union) for sending telegrams back and forth.[15]

In New York City, many such messengers worked downtown on Wall Street during the day. In 1883, the *New York Times* reported that "the members of the Stock Exchange and Wall-street speculators control the movements of a majority of the messengers absolutely," calling for boys from private homes and places of entertainment as well as from offices. Some firms would call for eight to ten boys first thing in the morning and keep them all day. The *Times* assumed that this was because there was not enough work to engage a "regular office boy," but clearly if ten messengers were being employed daily, enough work was available for a regular employee; instead, at 30¢ an hour, plus their free telegram-handling service, messengers were simply cheaper. Without intending it—and certainly without realizing it as such later—telegraph managers had created the first "temporary office labor" employment category over fifty years before firms like Manpower Inc. would claim the innovation as their own in the late 1940s (under strikingly different postwar economic conditions, feminized labor market demographics, and mechanized office information procedures, to be sure).[16]

Because of the urban markets for this kind of additional labor, the transformation of Western Union messengers into ADT messengers reinforced the

Table 3.1. *Top sites of telegraph messenger employment, 1930*

City	Average number of messengers employed at any one time
New York	2,082
Chicago	960
Philadelphia	444
Detroit	294
Boston	264
Los Angeles	261
St. Louis	244
Cleveland	232
New Orleans	204
Washington	197

Source: U.S. Census (1930).

urban concentration of messenger work. Through the first half of the twentieth century, from 1910 to 1950, New York City alone housed around 2,000 of the nation's 10,000 or more messengers—the largest number by far for any single urban area. Complete census data for messengers by city is only available for 1930, but that year gives a good example of where the top labor markets for messengers were located. New York, with its 2,082 messengers, employed as many boys as the next four cities combined—Chicago, Philadelphia, Detroit, and Boston. Altogether, the top ten cities alone employed some 43 percent of all U.S. messengers in 1930[17] (see table 3.1).

This new geography also affected messenger demographics. Multipurpose messengers were required to navigate the spaces of business quickly and unobtrusively, and for most of the twentieth century in the United States, such freedom of movement has been linked to class, race, and ethnicity (not to mention gender, which is discussed in chapter 6). As skilled craft apprenticeship waned and white-collar office work grew, the class differences among urban children grew more stark at the turn of the century: children of families of means (often headed by skilled craft workers or educated office workers) were kept in school in anticipation of better white-collar jobs upon graduation; children of poorer families (often headed by unskilled manual workers and recent immigrants), now living at home instead of boarding out with a craftsman, were under more pressure to contribute to the family income through waged labor. Such patterns had a gendered aspect as well, with the boys from poorer families more likely to work than the girls.[18]

Specific income and family data for telegraph messengers is lacking, but in

1907 the Bureau of the Census issued *Bulletin 69: Child Labor in the United States*, based on unpublished data from the 1900 census dealing with selected occupations and workers aged ten to fifteen. In one study of the broad "messengers and errand and office boys" category (which included more than just telegraph messengers, counting some 71,000 workers aged ten and over) a sample of 6,400 families of these workers from Boston, Chicago, New York City, and Washington was scrutinized in depth. Roughly 73 percent of the "messengers" lived with both parents, contradicting the myth that messengers were "orphans" or "sons of widows." Yet 57 percent of the children in these 6,400 families worked as "breadwinners." The investigators concluded that, "the pressure of economic necessity is very great in the class of families from which messengers and errand and office boys are recruited."[19]

In any case, the notion of "class" in U.S. society has always comprised more than simply a quantitative measurement of worker wages or family incomes. For example, class depends on the social relationship between worker and employer; thus, for ADT messengers, low piece wages in a subcontracted and contingent job were a powerful working-class marker. Class also depends on the perceived status associated with job tasks and work context. The telegraph industry was seen as a new and exciting technological field, part and parcel of the rise of "white-collar" office work in large, nationwide corporations; but at the same time, messenger labor within this novel context was seen as an old-fashioned, simple, unskilled service occupation. Service jobs bring another component into the class definition: the class position of the worker depends on the class position of the consumer. As telegraph transmission remained a rather high-priced service used only sparingly by those without means, messengers were associated with other servants of the affluent—hotel boys, elevator boys, house boys, errand boys, and office boys. Yet even with all of these class markers identifying messenger boys as members of the poorer classes, there remained a hope of advancement into either telegraphy or business that paradoxically added a sort of Horatio Alger legitimacy to the occupation.[20] (This is discussed further in chapter 8.)

As for "race" and ethnicity—that is, physical, cultural, and language characteristics and the prejudices associated with them—the messenger force was not completely homogeneous, because of the demands of telegram delivery. Although for many years WU had an unwritten policy of discrimination against hiring people of color and non-native whites—not a single "colored" operator was employed in the United States in the late 1880s, according to one WU manager—the telegraph certainly relied upon these diverse communities as customers. Because of the national scope of the network and the decentralized control of local managers, individual messenger forces matched their local labor and customer markets to some degree. In 1920, for example, *Telegraph Age* commented on the various ethnicities of messenger boys around the country: "in Boston, Timmy Dolan; in New York, Toni Caruso; in Milwaukee, Hans Shultz; in Minneapolis, Hans Swanson; in New Orleans, Poly Levine; and in odd places in California, Lee Wong, wear the telegraph company's uniform and generally remember to ask, in fair American, 'Any answer?'" In an industry based on communication, it could hardly be otherwise—as Western Union proudly proclaimed in 1916, the four messengers from its Santa Barbara, California, office were together able to communicate in Spanish, Italian, Greek, and English.[21]

A portion of the messenger force thus reflected the diverse American cultural experience quite early. For example, African Americans could be messengers in certain southern towns (and in some northern cities like Chicago), delivering telegrams to "colored" customers (so indicated by a notation on the outside of the telegram envelope in the 1870s). Spanish-speaking boys were needed in southwestern states, as in the El Paso, Texas Western Union office, described in 1912 as "unique in American telegraphic experience in that it is composed entirely of Mexican boys." And locals knew that delivering a message addressed simply to a "Mr. Cohen" in a Jewish neighborhood of New York City's Lower East Side could be futile unless a neighborhood boy made the rounds—perhaps one reason why by 1905 fully one-third of New York City's district messengers were said to be Jewish (a fact that allegedly caused the WU executive office considerable worry through the 1920s).[22]

The diversisty of local messenger forces resulted from both the demands of customer populations and the available supply of child workers, and eventually the telegraph companies began to advertise such situations in a positive light. In 1916, Western Union proudly pictured the twenty uniformed boys of "The Colored Messenger Force at Jacksonville FL" as "a credit to themselves and to their race." But in many places, such arrangements still generated problems. In 1903, *Telegraph Age* noted that "the question of color is asserting itself in the ranks of the boy messengers" in Macon, Georgia: "Because the service rendered by the white boys was not satisfactory to the [apparently black] residents of that town, and colored lads were put in their places by the employing company, the black force has been peppered with rocks by the displaced youngsters, whenever the former have emerged upon the streets bearing the yellow envelopes." No explanation for the "unsatisfactory" service of the white messengers was offered—either the preferences of the black customers or the bigotry of the white messengers (or both) could have been at work. One former (white) messenger recalled a similar situation in 1946 that, fortunately, had a more benign ending: "We had a 'Little Harlem' area in one town where white boys weren't wanted. The first few times I made deliveries there I was scared. Once they

Figure 3.3. *U.S. telegraph messengers by race, 1910–1950*

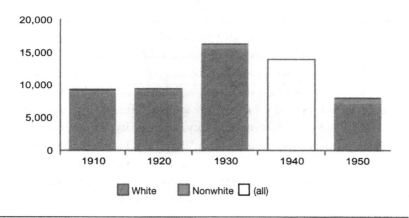

Source: U.S. Census (1910–50).
Note: No breakdowns by race for 1940.

came to know me, they would yell, 'Here comes the telegram boy' and even help me locate people in the area. It was a real relief."[23]

Either way, such conflicts were rare in the overall telegraph network, because even with all those necessary pockets of diversity, throughout most of the United States in the early twentieth century official counts showed telegraph messengers to be overwhelmingly "white" (97 percent in each census from 1910 to 1930) and, specifically, overwhelmingly "native" white (from 87–92 percent in each of the three censuses). This category did include many second-generation European immigrants, who would have been able to maintain a link to first-generation immigrant neighborhoods. And certainly many culturally diverse messengers may have been missed by these counts altogether. But it seems that the proportion of white telegraph messengers was significantly greater than that of white workers in the labor force as a whole in 1920 and 1930, when the Census counted that whites made up only 88 percent of all civilian workers[24] (see figure 3.3).

The national/local division in scale between WU and ADT served to hide such divisions in messenger labor and prospects. Local ADT offices, not officially part of Western Union, could and did employ boys of different backgrounds as their local customer populations demanded (and as their local labor markets allowed). But this diverse army of messengers at the local fringes of the telegraph network never threatened to rise into positions of prominence in the national system; after all, Western Union did not even employ ADT messengers, but only subcontracted them. Thus, the growing number of ADT boys

lacked the opportunities to learn telegraphy that their Western Union prede-
cessors had enjoyed (as described in chapter 2).

Messengers were quick to recognize this blocked career path, and messenger turnover became an increasingly serious problem for ADT managers. As a
result, the company became more and more careful about hiring the "correct"
boys. In 1875, for example, prospective messengers had to sign an employment
contract with ADT giving their name, age, birthplace, residence, place last
employed, references, and a handwriting sample. Often their parents had to
sign as well. By 1879, ADT New York City had set up its own messenger training school, in which boys learned not only "the general business of the company,
calculation of tariffs, etc.," but also later "all known tricks of confidence men
and sharpers," "the routes and streets and railroads of the city and vicinity," and
"how to behave." ADT hoped that such training would weed out problem messengers early.[25]

Training schools served another purpose by stringing boys along without
pay until they were actually needed by the district companies. The ADT
Philadelphia managers were proud of their 1878 training school because "boys
receive no pay except when doing actual service, thus enabling us to keep a force
in reserve with little expense and meet any sudden or fluctuating business to
which we are especially liable." Furthermore, they were able to "reduce our regular force of boys to a point below any heretofore reached with safety and at the
same time meet the demands of our business." In this way, training schools were
yet another technological load-leveling strategy to make efficient use of the idle
time of the messenger boys (as discussed in chapter 2).[26]

Despite such efforts, messenger turnover continued. Resignations and firings resulted in yearly turnover rates of 100 to 300 percent. For example, in
1875 at ADT New York City, 1,334 messengers were hired to fill an average
messenger force of 427 (116 of whom were "officers" who didn't themselves
perform regular messenger work). ADT Philadelphia didn't expect messengers
to stay very long in 1880: "Messengers as they advance in years are obliged to
leave us to learn trades or enter into some business that promises permanency
in the future." That year, to maintain a force of 215 messengers, 220 boys were
"discharged or resigned"—the "training school" notwithstanding. Five years
later, the situation was worse: to maintain an average of 203 messengers on
duty, 529 were discharged or resigned. The company explained this turnover
using two arguments: "1. Good boys resign because someone else wants them
and offers better wages. 2. Bad boys are discharged because we don't want them
at any price." Thus, it would appear that ADT's remaining boys were not good
enough to get hired elsewhere, but not bad enough to be fired.[27]

High messenger turnover was more than simply an annoyance. In the new
dual-scale telegraph network, comprising both intercity and intracity systems,

competition was fundamentally altered. Space, technology, and labor were all factors in the competitive battles that WU fought with other telegraphs during the last years of the nineteenth century. The more the national telegraphs relied on local district companies to gather and disseminate their messages, the more these local district telegraphs were the sites of conflict between competing national firms. And the duties performed by district messengers, not to mention the wages and circumstances of their work—including turnover—were important levers in such competition.

As early as 1871, when the rival Atlantic and Pacific Telegraph Company (A&P) was emerging as a competitor to WU on the East Coast, WU had considered lowering their telegram prices. But this would have cut revenue drastically and spawned a "rate war"; instead, WU instituted free telegram delivery in New York City (through messengers) to better compete with their new rival. WU's deal with ADT two years later, seen in this light, was a way to preserve their competitive position against the A&P by both improving delivery speeds and giving customers an easier way to choose WU as their telegraph provider.[28]

Under "robber baron" and future WU owner Jay Gould, the A&P took this sort of competition one step further. In 1876, Gould's A&P office in New Haven, Connecticut obtained a list of local WU customers, and offered to provide them all with ADT call-boxes at a lower rate than WU was charging, in order to capture the call-box telegram revenue from WU. Western Union responded by purchasing all those call-boxes outright from ADT at $5 each plus a $1-per-year fee, so that the instruments could only legally be used to call for WU telegrams. Western Union estimated that it would reap at least $12 per year from each call-box customer in telegram revenues, so the $6 first-year cost and subsequent $1-per-year maintenance cost seemed reasonable. Thus one of WU's main competitive strategies in the late 1870s was to strengthen its contracts with local ADT systems and to purchase, if necessary, local ADT technology. In 1877, Western Union's new contract with its ADT licensees in the eastern and central divisions (including New York City and Chicago) gave WU exclusive rights to manufacture and use ADT call-boxes in order to "prevent the American District system from falling into the hands of the opposition."[29]

Though WU eventually ended this battle by simply buying out Gould's A&P (at a tidy profit to Gould), the episode was merely the first act of Gould's Western Union drama. In the late 1870s, Gould formed another telegraph company, the American Union Telegraph Company (AU), by buying up smaller concerns, this time helped from the "inside" by WU's own Thomas T. Eckert (of Civil War fame). Again, in its most important markets, this competition led WU to strengthen its deals with the district companies. In August 1880, ADT New York City entered into a twelve-year agreement with WU to collect and deliver telegrams in New York City. As the year went on, WU set up similar relationships with independent ADTs in other cities, sometimes buying the district companies outright. Worrying that the "Opposition Co." was putting in its own "call-bells" to woo WU customers away, WU considered buying up the entire capital stock of ADT Baltimore, ADT Dayton, and ADT Louisville. This was a shift from two years before, when WU was content to merely arrange contracts for service with local ADTs.[30]

Individual ADT franchises faced new competition as well. After complaints about "inefficient service" by brokers at ADT New York City, in April 1881 the new Mutual District Messenger Company placed its own call-boxes free of charge in Wall Street offices, and offered its messengers a percentage on any messages they carried, in addition to their salaries. Now brokers could call for messengers from both companies at once, giving the job to whoever got there first. This was when ADT salaries in New York City were $3.60 per week for twelve-hour days, so the "commission" offered by Mutual couldn't have been very large (and in any case, commissions were one step closer to a piece wage). ADT was eventually able to buy out Mutual years later, but the competition came at precisely the worst time for WU.[31]

Waged through message pricing and district service, this time Gould's war with WU had a different effect: after using his own AU telegraph as a lever to drive the stock price of WU down, in January 1881 Gould sold his Union Pacific railroad stock high and bought 90,200 shares of WU stock low, enough for AU to take majority control of WU. At the news of the merger, both WU and AU stock rose in value, while public outrage at the new "telegraph monopoly" mounted. Gould remained in control of WU, and named his ally Eckert as WU general manager. WU's former backer William Vanderbilt was voted out of the board, and Eckert was voted in. The Gould family would retain stock control of WU until 1909, when AT&T would take over the company.[32]

Whether one believes that Gould "wrecked" WU or that he saved it, the immediate result of the takeover was the organization of a slew of other telegraph companies to battle WU, riding a wave of anti-monopoly sentiment: B&O Telegraph, Mutual Union, Bankers and Merchants, and American Rapid, to name a few. Gould's WU tried to buy up control where it could, such as with Mutual Union, and by 1884 WU handled 92 percent of all U.S. telegraph business, with 14,069 offices and 20,055 employees.[33]

Though WU was nearly a monopoly, there was one company that would remain to challenge it well into the twentieth century. In 1881, the Postal-Telegraph (PT) company was created, so named to suggest that the U.S. post office patronize this new firm exclusively. Ironically, the relationship ended up working the other way around: PT had difficulty breaking WU's exclusive contracts with railroads for trackside right-of-way, and had to run its lines along postal roads instead of along rails. Even as late as 1922, PT only linked "the big cities and towns of medium size," handing over "messages for the backwoods" to the mails.[34]

Postal Telegraph also lacked call-box circuits, even in New York City. Their solution was to send messengers around to big companies at specified intervals during the day to pick up any business that had accumulated. If no messengers were available at the appointed time, the manager himself would go on these rounds, leaving the office empty, indicating that the collection and distribution of messages for these large customers was of paramount importance—and that messengers were critical.[35]

Postal's early years were tumultuous. Only two years old, the company was bought by John W. Mackay, a man who had made a fortune in the Comstock Lode silver mine and then used it to lay a transatlantic telegraph cable. Mackay reorganized the company as the Postal Telegraph Cable Company; a year later, the company was reorganized once again after defaulting on its bonds. Postal Telegraph ended up as an association of fifty or so smaller companies, but it survived, and even innovated—in 1903, PT's Clarence H. Mackay, J. W. Mackay's son, completed the first transpacific telegraph cable.[36]

PT tried hard to catch up to its rival. Just as WU had invested more and more resources into its relationship with ADT, PT finally set up its own district messenger service when it incorporated the Postal District Messenger Company as a New Jersey holding company in 1901. But the two companies were more than simply competitors. As early as 1888, WU and PT reached an "understanding" to end the practice of giving rebates and special low message rates to big commercial customers in order to woo them from each other, according to former PT president A. B. Chandler. Such a move served Gould's interest to keep PT alive as "competition," avoiding charges of a WU telegraph monopoly. But it runs contrary to the usual story of competition through capital investment in new technology. By the late 1800s, capital costs for the telegraph industry were minimal; instead, labor costs were the main drain, and thus the main source of both competition and cooperation.[37]

Consider the actual costs WU paid for setting up offices in the 1880s. A new office in Champaign, Illinois, demanded a capital investment of $135 (extending lines and "fitting up office") and monthly operating costs of $95 per month. The operating costs included $25 for rent, light, and fuel; $60 for the operator's salary; and $10 for the messenger's salary. The total fixed costs of setting up the office were equal to two months of labor costs, and nearly three-quarters of the monthly costs were labor costs.[38]

Such labor costs were themselves uneven; while operator wages were relatively high, messenger wages were very low. Even at ADT New York City, a company built on messenger boys, messenger labor costs in 1880 only made up a little more than one-third of all company expenses. Western Union's in-house

Table 3.2. *ADT New York City receipts and expenses, 1879–1881*

Year	Gross receipts ($)	Gross expenses ($)	Messenger wages ($)	Gross expenses (%)	Net revenue ($)
1879	364,800	266,800	90,324	34	98,000
1880	442,839	318,584	117,000	37	124,255
1881	467,948	380,207	156,000	41	87,741

Sources: *Operator* (1881); *NYT* (1880–81).

(non-ADT) messenger wages from 1878 to 1880 ranged from $2.50 per month to $20 per month, with most coming in at $8 to $12 per month. The average of 106 cases over the two years was a messenger wage of $9.75 per month, or about $2.50 per week[39] (see table 3.2).

Low wages were coupled with long hours. In 1887, WU messengers in New York City claimed that at a piece rate of 2¢ per message, it was hard to make even $1 per day, especially since "The company frequently wants us to sleep on benches in the office, when we get in very late, and start right in again to work in the morning." American District Telegraph messengers at this time worked over twelve hours a day for a wage of $3 to $4.50 per week and only received 5¢ per hour of overtime. Postal Telegraph messengers weren't any better off: one former messenger recalled that he started in the 1890s earning $12 per month for work hours from 7:30 A.M. to 8:00 P.M. with a half hour for lunch, making a twelve-hour day, seven days a week.[40]

Thus, messengers were crucial to telegraph competition, but were incidental to the wage budget. Western Union president Norvin Green reported that in 1888, the total revenue to WU from the 51.4 million land messages it sent was 31.2¢ per message, with each message costing the company 23.2¢ (and with WU making 8¢ profit on each message, or 25 percent). Where was the 23¢ spent? Green said 13¢ went to salaries for operators, clerks, managers, and officials; a little over 4¢ went for maintaining lines; and 2¢ went for messenger delivery (the messenger piece wage). The remaining 4¢ went to batteries, fuel, rents, and the like. In statistical notebooks prepared for WU president Robert Clowry, the same pattern appears from 1893 to 1908. Nonmessenger salaries made up between 41 and 44 percent of total expenses, but messenger salaries only made up between 7%–8% of total expenses, or one-fifth to one-sixth of the wage bill for the rest of WU's employees.[41]

Both WU and PT tried various tricks to extract more labor from their operators for the same 13¢ per message. As early as 1868, the "sliding scale" was used by WU managers to take advantage of the geographic unevenness in office salaries. Under this rule, when operators switched offices, they took the lower of either the salary from the old office or the salary of the new office as their new wage. This greatly affected so-called boomer operators, who would travel from city to city, working just long enough in each place to earn enough to keep traveling. Technology played a part as well: the duplex telegraph system developed in 1872, which sent two telegraph messages along a single wire simultaneously, increased message capacity and motivated an operator speed-up. Increased Morse speed can only come through increased operator skill, a skill for which operators often demanded increased payment. But the duplex (and later the four-message-at-once quadruplex) was only used on high-traffic trunk lines, adding another layer of geographic unevenness to the salary mix.[42]

In 1902, WU changed its payday to bimonthly instead of weekly, reduced operator salaries by 10¢ per hour, and transferred more and more operators to the "extra list" or "waiting list," meaning operators had to show up for work and be present all day, without any guarantee that they would *get* any work that day (and they were only paid for hours they actually worked). A "bonus" system was used to speed up operators: after 325 messages in one day, every extra message an operator handled was worth an extra 1¢. The "split shift" (also called a "split tour" or "split trick") had operators working an eight-hour day but with a four-hour break in between. This kept them working during the busiest times of the day; it also kept them occupied for twelve hours while they were only paid for eight.[43]

While operator wage and hour schemes grew more complex, messenger compensation was simplified into a piece wage, for several reasons. First of all, the piece wage provided a mechanism for the manager to pay only for labor (actual work performed) as opposed to labor power (the potential to work during a given time, such as a day). Second, through setting and resetting the piece wage, managers could also increase the pace of work. And third, the piece wage served as a convenient load-leveling strategy for telegraph managers. In 1884, ADT Philadelphia managers lamented that even though their messengers worked eleven hours a day for their wage, "As business fluctuates daily, it is not possible that every boy can be earning money every moment he is on duty. There is therefore always a large margin of expense that does not bring a corresponding profit to the company, but is an actual loss that cannot be obviated from the peculiar nature of the business itself." The piece wage ended this "actual loss."[44]

This worry over wasted messenger time was an ongoing concern, especially in large cities with vast but uneven messenger staffing needs. As late as 1926, WU appropriated $1,375 for tabulating a study of 400,000 delivery and pick-up runs "on which messenger compensation is based." Four months later they committed another $588 to the effort. This was not a great deal of money to WU by any means, but the study illustrates the continuing quest for firm data on which to base messenger wages and work rules.[45]

Load-leveling took other forms as well. In New York City in the early 1920s, WU set up a central telephone switchboard where branch offices requested extra messengers on a day-to-day basis—"waybills" loaned from one office to another. By 1934, messenger allocation in New York City was centralized and rationalized, with four men seated around a table wearing headphones, working at the New York City "Messenger Dispatching Center" as they shuffled idle messengers among two hundred branch offices to balance the load during the day. In the most extreme example, on December 24 each year, eight hundred downtown messengers were shifted to midtown offices, to run holiday errands for florists, confectioners, and department stores, since the financial business only ran for a half day. In the same way that an automatic multiplexer shuttled the electrical signals of four telegrams simultaneously over a single wire, the messenger dispatchers allocated and reallocated the boys ceaselessly, searching for maximum throughput. The overriding metaphor for allocating personnel was a mechanical one: "When trouble does occur, standardization and interchangeability of men, methods, and machines, helps the organization to 'get the circuit going' and then repair the failed elements."[46]

From the point of view of the messengers, such "scientific" distributions of labor and wages resulted in a wide variability in what each boy actually earned. A messenger's weekly wage depended on how many hours he was allowed to work, whether he worked during the busy times of day or the slow times, whether he was on hand for the lucrative holidays that made up more and more of the telegraph's profit, and what his relationship was with his supervisor (office manager or delivery clerk). The piece wage could be lucrative for a boy who got the right route in the right place at the right time; otherwise, the piece wage could be a pittance.

For example, in the 1910s and 1920s, large WU customers often sent what were known as "books" of telegrams. Each book consisted of a "form-letter" telegram to be forwarded to everyone on a list of recipients—anywhere from twenty-five to a thousand addresses at once. Each copy sent was considered a single telegram, both in terms of cost to the consumer and in terms of payment to the messenger. One former messenger remembered, "Picking up such a 'book' was a huge windfall for a messenger; he might earn ten or fifteen more dollars, 2-1/2 cents for each additional name on the list, if the list was long." This boy was lucky because "quite a number of secretaries seemed to save such 'books' for me," such that "I was offered several jobs that seemed promising, but did not pay nearly what I was soon earning as a messenger, fifty dollars and more in a week—more than the office manager was getting." Of course, the more lucrative the messenger gain, the more incentive there was for the customer to avoid using the messenger. In 1925 in a small-town Tennessee office, even though the local flour mill regularly sent books of twenty-five to thirty telegrams, these books weren't handled by messenger, but by a mill employee—so no one claimed the windfall.[47]

If the "book" was the messenger's dream, the "soak" was the messenger's nightmare. This was a single message that took a long time to deliver because it was far away. A soak was usually a delivery to a distant private residence, such as a death announcement that couldn't be phoned. One former messenger recalled with disgust the soaks he was forced to deliver in 1935—messages that took an hour or more to deliver, taking up the middle of the day, but that still only earned the messenger 3¢ apiece. The pain of the soak message was at least offset a bit by a "zone system" that paid a slightly higher rate for distant deliveries.[48]

Between the books and the soaks were the vast bulk of messages where the only hope for something extra was to convince the recipient to tip generously. Officially, WU discouraged tipping: "We would prefer that our patrons do not tip the messengers for service which we pay them to perform," said one manager. But customers didn't follow the WU rulebook, and messengers quickly learned which routes earned them the "overs" and which routes did not.[49]

Here the dynamic of profitable business books and unprofitable domestic soaks was reversed: businesses rarely tipped, but domestic residences might. Tips in New York City came not from Park Avenue or Fifth Avenue (where servants often received the messages), but from the Lower East Side: "Frequently the man is unable to read and write; the boy is called on to read the message to him, and then to write out his reply, and all these services the East Side patron thinks he must pay extra for." Though a bit romantic, this description does agree with other accounts. One messenger remembered that in 1921 in New York City, when delivering telegrams in the uptown districts of "imposing mansions," "almost always a servant would accept and sign for the telegram, often without even a 'Thank you!'" However, at the smaller houses, "no matter how small or apparently poor, there was ten or fifteen cents, and many times a quarter, as a tip."[50]

Another way a messenger might profit from the piece wage was to work the busiest days of the year: holidays. Messengers who worked on Christmas, New Year's, or Mother's Day could sometimes expect a bonus, in addition to the large volume of messages they carried. In 1925, WU approved double-time pay for delivery department employees who had to work on Christmas, costing the company a total of $6,500 for the whole department. But conditions during the holidays were grueling, as messengers described to Congress in 1938: "Boys as young as 16 years of age were working 18 and 20 hours a day during this 3-day period," and "received no lunch or rest periods." A former WU New York City night bicycle messenger said quotas were set for messengers on holidays, and "it is insinuated that if they fail to get this quota they will either [be] discriminated against or transferred."[51]

Still, messengers were motivated to work holiday shifts because for a few days, tips were plentiful. Each Christmas, the editor of the telegraph journal *Operator* printed up and sold holiday greeting cards that messengers were supposed to purchase and then hand out to their customers, hoping for a tip in return. The 1880 card read, in part:

Another year has passed away
Since you gladdened our hearts one Christmas day,
Causing them to beat with heartfelt joy,
As you kindly remembered the telegraph boy.

While you sit by your warm fires,
We bring you the news from our great wires.
It rains, it hails, it sometimes snows;
But we carry the message where'er it is to go.

Such holiday solicitations, while making the messenger into more of a petty capitalist, could backfire as well. In 1885, two former WU messengers in New York City borrowed uniforms from friends still in the service and collected over $500 from seventy-six downtown businesses by soliciting donations with a "poetical circular." Such scams were legendary; from 1923 to 1930, the *New York Times* ran warning blurbs each holiday season, saying that patrons should not give to any "messenger holiday fund" because such a charity did not officially exist. Middle-class fears of beggars posing as messengers worked against the efforts of the messengers themselves.[52]

Though crucially important to the messengers, their wages, whether earned by the piece through the delivery of messages or earned on the side through the occasional tip, were only half of the story. Wages could be taken away as easily as they were handed out. A system of fines was often used to enforce day-to-day messenger rules. Boys were to "Keep the uniform, shoes, hands, and face clean, and the hair cut frequently," and "Be orderly and quiet while in the office awaiting turn to deliver messages." In 1887, ADT messengers were subject to fines such as 30¢ for "wasted time" and 50¢ for slipping a message under a door instead of delivering it in person; they could be fined three hours' wages for being five minutes late, or they could be compelled to work more hours. The rationalized system of penalties was designed to set a value on every minute of the messenger's labor.[53]

How did a messenger avoid the soaks while cornering the books and the high-tipping deliveries? How did he make sure that he was on duty for the best hours of the best days? How did he keep from being fined by an angry or corrupt supervisor? In large offices, kickbacks to delivery clerks and office managers could earn messengers special treatment. One messenger who worked on the Magnetic Telegraph line in 1849 earned $2 per week as one of a dozen boys, all aged twelve to sixteen, who served the New York area. His manager "levied a percentage on all 'extras' made by the boys," and "woe to the boy who attempted any 'shenanigan' with the old man, as he had the names of all who paid, and the boy who failed to make a fair division suffered for it by being cut off from the delivery of messages that yielded a gratuity." Similarly, WU messengers in the 1880s complained of rampant favoritism, where one boy would be given many nearby messages all at once while other boys would have to deliver distant messages one at a time—with all boys earning the same fee per message. ADT messengers argued that bribing the manager for a good route under this system could be worth an extra $10 to $12 a week. And such favoritism was inevitably tied to the prejudices of the office manager. Popular author Harry Golden, reminiscing about his experiences as a Postal messenger in New York City in 1913 (when he was eleven years old), remembered that the coveted hundred-telegram books were given out not to Jewish boys like him, but only to Irish and Italian boys. Such were the unwritten rules of the messenger service, based on both the decentralized control of offices and the unevenness of piece wages.[54]

As these last two chapters have shown, the growth of the telegraph system in the late nineteenth century was a mixture of three stories. One was a technological story of wires extending along railways and roads, spreading across the states and territories through both civil war and westward enclosure. Impulses traveled along these wires first one at a time, then bidirectionally, and finally in groups of four or even sixteen, as the speed of human fingers (and of human attention) gave way to machine-mediated typewriters and transcribers. In this story the Euclidian space of the nation was wired, and the cyberspace of (theoretically) instantaneous communication was created.

A second story centered around the business of the telegraph, full of inventors, entrepreneurs, and controlling stock interests. Telegraph companies started locally, but lived or died on the cooperation they found from larger regional interests such as railroads. Through both wild speculation and methodical consolidation, it was Western Union, under different factions of financial control at one time or another, that emerged as the nation's telegraph company, though it entered the twentieth century with the smaller Postal Telegraph nipping at its heels. Both of these companies relied on subcontracting relationships with the district telegraphs, the wirers of cities and employers of messengers that funneled the telegrams into and out of the national networks. The national space of telegraph command and control was founded on local spaces of competition.

A third story can be found between these two, in the spaces of labor. While adult male Morse operators and female Automatic operators were confined inside on several huge, noisy floors of the Western Union headquarters in New York City, young male messengers roamed the streets and braved the traffic, funneling messages into the building and then carrying them out. The telegraph system thus existed on two spatial scales at once, with intercity messages flowing on railway wires, and intracity messages flowing on scraps of paper carried under the hats of the messenger boys. The next chapter will begin to consider these urban spaces of messenger labor more closely.

THE SPACES OF MESSENGER DISCIPLINE

The messenger's work is light, interesting and important. Never confining or irksome, it keeps him much of the time in the open air.

—Western Union messenger manual, 1946[1]

It was a slaughterhouse, so help me God. The thing was senseless from the bottom up. A waste of men, material, effort. A hideous farce against a backdrop of sweat and misery.

—Henry Miller, former WU messenger manager, 1920s[2]

Just as the messenger's wages and duties were set early in the history of telegraphy, so were the measures of messenger discipline. In 1850, new Magnetic Telegraph president William Swain "ordered that a messenger should not be allowed to take out more than one message at a time, and that he must be timed, and obtain a timed receipt." Swain allegedly proclaimed, "Boys must wait for messages, not messages for boys," making Magnetic the model "for regularity, promptness, and dispatch." By the time of its monopoly merger in 1866, Western Union agreed that effective messenger service was crucial, emphasizing that "*The success of the business*, and the credit of the Company largely depend upon the *promptness with which the business is done*, and no branch of it requires greater energy, care and promptitude, than the *delivery of messages*." But WU focused on the manager, not the messenger: "Especial [*sic*] care and watchfulness on the part of each Manager is required against delay in this respect."[3]

A decade later, ADT faced such questions as well. In 1875, district managers were instructed to file weekly reports on their messengers, ranking each on seven-point scales of "promptness, conduct, industry, obedience, cleanliness and dress" that provided grounds for firing if unacceptable. But in 1878, ADT Philadelphia officials admitted that managing their messengers had become such a "source of trouble and vexation" that they were forced to hire a messenger superintendent in order to "bring the boys up to that standard required to bring about success." By 1879, they even employed a "messenger detective" to "watch and follow the messenger in the performance of his service," surveillance that reportedly had "a very beneficial effect upon [a messenger's] habits, promptness and appearance." Yet things were still so bad a few years later that in 1885 the *Electrical World* joked, "in some particulars, the dog would certainly be an improvement on the messenger boy. Dogs do not play marbles, nor will they go many blocks out of their way to follow a circus."[4]

Managerial efforts to control labor are part and parcel of capitalism. What is striking is not the existence of discipline techniques for messengers, but the form that such discipline took. Besides the general temporal discipline of the telegraph labor force (with transit and return times recorded and closely monitored), messengers were subject to two other forms of discipline: the quasi-parental discipline thought necessary for young boys, forever yearning to play instead of work, and the spatial discipline thought necessary for employees working outside of the telegraph office, out of direct sight of supervisors. The fact that messenger employment was largely an urban phenomenon meant that in order to produce and control its messenger force, the telegraph industry had to also produce and control urban space itself. Such measures acted at three different spatial scales at once: the indoor realm of the telegraph office, the outdoor realm of the urban street, and the very body of the telegraph messenger himself.

The messenger's main function in the telegraph industry was always the delivery of telegrams, so it is not surprising that the telegraph managers attempted to precisely script the way deliveries were handled. After all, they reasoned, the job was an unskilled one that could be learned in a day, if one followed the rules. As illustrated in chapters 2 and 3, written codes of conduct, standardized piece wages, and financial penalties were all used to micromanage the delivery of telegrams, with respect to both space and time. Spatially, message delivery was a task that occurred outside of the telegraph office itself, and was thus fundamentally *un*manageable. Temporally, the speed of message delivery not only determined how much of the "lightning speed" of telegram transmission would be passed on to the consumer, but also how quickly a messenger could return to the office, ready for another run, removing the need to hire more boys.

But once ADT emerged in the 1870s and gave messengers a whole new range of duties, from package delivery to part-time employment, new practices were needed to properly control the boys. Through mechanical and military metaphors, the ADT boys became numbered components in the telegraph system itself, and the messenger rules became militarized codes of youthful conduct, resulting in a uniformed messenger army that was soon copied by all the other telegraph companies.[5]

The first step was a simple but pervasive one: in larger offices, instead of referring to the boys by name, they were each assigned a number. This small change in record-keeping made sense in an environment where job turnover was so high that "Messenger 13" might change face regularly from week to week. But the rule led to even more changes that show how a mechanical metaphor was being applied to these human telegraph components. For example, in larger offices, messengers were tidily arranged by number just as one might line up machines or bundle wires. As early as 1869, WU's Chicago office bragged, "Seats are provided for these boys, and ranged along at a convenient point, each seat having a certain number, by which its occupant shall be known. The boys will have no other name than that of the chair they occupy, save on the pay-rolls of the Company." By 1910, the official WU messenger manual likened to the messenger boy to a "cog in the wheels of a big machine."[6]

ADT transformed the mechanical metaphor into a military one, with the addition of drills, ranks, and uniforms. The Civil War had put a generation in uniform, whether blue or grey; the business depressions and strikes from the 1870s to the 1890s spawned a National Guard of uniformed "peace-keepers" working as hired hands to capital; and a new generation of educators would stress military precision in the schools well into the Progressive period. But the military metaphor worked in other ways for the telegraph companies too, especially in a "private utility" constantly threatened with government takeover.

The messenger uniform was the most important change of this kind. Uniforms served many purposes to ADT managers, advertising messenger and telegraph service to potential customers while at the same time unmistakably marking messengers (and their behavior) to observers in the crowded city streets, acting as a form of automatic, silent discipline (insurance against both tardiness and mischief). On the other hand, to messengers this same marking carried benefits, as the uniform was a sort of "press pass" allowing them access to the halls of commerce, government, and entertainment. Nevertheless, the cost of renting uniforms and the labor involved in keeping them clean (a duty that WU relegated to the mothers of the messengers) were significant.[7]

The precise origin of the messenger uniform is unclear. Just as the messenger's duties echoed those of the postal letter carrier, and their wages compared to the 2¢-per-letter urban delivery rate, so may the messenger uniform have been modeled on the post office as well. In 1868, only a few years before ADT was formed, Congress and the postmaster general mandated that all letter carriers wear uniforms. A more direct antecedent came a few years later from England. In 1870, six hundred telegraph messengers in London, aged thirteen to fifteen, were organized into drill corps, given uniforms, and told that good performance would lead to promotion: "It is of little avail that a message should travel at lightning speed from one end of Great Britain to the other if the messenger charged with delivering it dawdles over his work," said the *European Mail*. A year later, the *Telegrapher* urged a similar practice for U.S. messengers, saying it would help prevent other boys from posing as messengers and swindling the public, if the uniforms were kept by the company so they couldn't be worn off-hours. Thus from a badge of civil-service pride in London (where telegraph service had been nationalized into the post office in the 1860s) the uniform was translated across the Atlantic as a tool against graft in the world of private commerce.[8]

No matter where the inspiration came from, the uniform changed the messenger relationship, inaugurating legal contracts with every boy. At ADT St. Louis in 1874, each messenger had to sign a form that read, in part:

> The undersigned having this day commenced service as a Messenger of the AMERICAN DISCTRICT TELEGRAPH CO., at a salary of *$3.50* Dollars per week of seven days, (day's work, 10 hours,) hereby agrees to purchase at the current prices, a Uniform, (Coat and Pantaloons,) Rubber Coat and Leggin[g]s; also Cap and Cap-cover; the same to be paid for as follows: 1st week, *$3.50* and each succeeding week, $1.50 until articles are paid for. . . .
>
> It is further understood, that the "stripes" and buttons on the Uniform, the badge upon the cap, and the shield worn upon the coat, being the peculiar insignia of this Company, it is therefore agreed, that when the undersigned quits the service, he shall return the aforesaid articles without compensation or equivalent therefor.

The filled-in salary and uniform cost reveal that ADT could raise the price of the uniform according to local conditions. In this example, the boys worked the first week unpaid, and then worked for $2 per week until their uniforms were paid for. There is no telling how long that might have taken, but if both summer and winter uniforms cost around $10 each, boys could easily work two months each year at the lower wage rate.[9]

At the start of 1874, WU borrowed the ADT uniform innovation as well, and all 130 WU New York City messengers began to wear uniforms: "dark blue pilot cloth, well made and warm, the surtout buttoning close up to the neck, with a leather belt round the waist, bearing the name of the company." Separate summer and winter wardrobes were created, and in 1877 WU first agreed to arrange for the manufacture of one hundred pairs of shoes for its New York

City messengers—to be resold to them at cost. In 1875, ADT even put its managers in uniforms of a sort: "Prince Albert coat; vest and pants, the rank being indicated by the color and arrangement of buttons." The buttons bore likenesses of messenger boys.[10]

It is difficult to gauge how the messengers reacted to the uniforms. Western Union messenger manager Mike Rivise claimed in 1950 that "The uniform was so much a part of the glory of being a messenger that the former A.D.T. boys, given a choice of wearing their own clothes or renting uniforms, *all* grudgingly paid 50¢ a week rental for uniforms out of their $3.50 weekly salary." But if many enjoyed the prestige the uniform implied, others felt it to be a ridiculous encumbrance. The *Telegrapher* suggested that the uniform was not welcomed easily by messengers, recalling that "a small rebellion was created among the limited number of messengers then employed, by an attempt which was made to have them wear a uniform cap" in the days before ADT.[11]

Clearly the uniforms cost money, both to the telegraph companies and to the messengers. ADT was forced to organize a "Bureau of Repairs" in order to maintain the uniforms for its more than five hundred messengers. By 1878, WU was buying uniforms in bulk: winter uniforms cost $10.50 each, caps a little over $1.10 each. Twice a year, in winter and summer, uniforms would be resold to messengers, but in the meantime they still had to be stored and maintained. By 1924, WU was spending $350,000 and up per year for messenger uniforms, and took out insurance on sites where they stored over $5,000 worth of uniforms together. The high point of $550,000 was spent in 1929, the year of the peak messenger numbers right before the Great Depression. But even through the Depression, WU often spent over $100,000 per year on uniforms[12] (see figure 4.1).

All of these uniform costs were to be passed along to the messengers. By 1882, Mutual District messengers in New York City had a total of $33 deducted yearly from their wages to pay for their uniforms, which consisted of pants and coat, a cap, rubber pants, and rubber coat (though the companies demanded that the uniform be visible, so the rubber pants and coat were often not worn). In 1885, the Mutual boys even held a "Minstrel Show," raising $75 to "procure uniforms for boys who would like to enter the messenger service, but who cannot afford to buy uniforms." In the 1890s, one former PT messenger recalled, boys had to purchase uniforms from the company costing $8.50 (payable over two months), caps costing 75¢, and badges "rented" for 75¢ and returned when the messenger quit or was fired. By taking such fees out of messenger salaries, the managers claimed that they were building responsibility, making sure the boys would take care of their uniforms; by not allowing messengers to buy badges outright, managers were keeping control over their brand image and message security.[13]

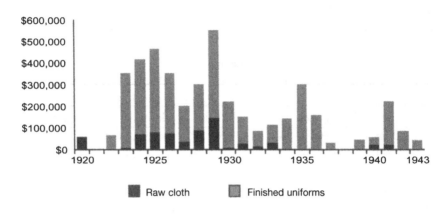

Figure 4.1. *Western Union messenger uniform expenses, 1920–1943*

Raw cloth ■   Finished uniforms ■

Source: WU, Executive Committee Meeting Minutes (1920–43). WUA, 1993 addendum, series B, boxes 8–30.
Note: Years with no figures may represent missing data.

Messenger reaction to the uniforms in the 1930s varied just as it did in the 1880s. One former messenger from a small Georgia city in 1934 remembered, "I was proud of the uniform. Also that uniform was our ticket to just about anything and anywhere free of charge. Circuses, fairs, traveling shows, rides, special events. Of course, we didn't get free admission to the local movies, but I can't remember ever being denied access anywhere or anytime. Yes we were allowed to wear our uniforms home and I did." But a different opinion was expressed at the 1938 Senate Committee on Interstate Commerce hearings on the telegraph: "[t]he companies have the brazenness, in some localities, to extort—and I use the word advisedly—to extort from the messengers a charge for the use of the uniforms which the messengers wear, which charge is about 6 cents per day, thereby taking from the boys about 50 cents a week, on the average, out of the $6 or $7 they receive, or, in a good week, $8." And more than that, "We have deplorable cases—and I am not attempting to shock you—where the uniforms are literally vermin-ridden. They were dilapidated; shirts with tails missing and sleeves missing; with big holes; trousers woefully stained and patched." This represents another geographical difference: in a small city with few messengers and low turnover, a clean uniform was expected, and was worn with pride; in a giant city with many messengers and endemic turnover, recycled and worn-out uniforms were the norm, and were worn only reluctantly.[14]

The uniform persisted even in the face of such criticism partly because it fit together so well with other forms of messenger discipline, such as Western Union's messenger numbering system (soon expressed on the uniform cap badge). In particular, messenger uniforms helped legitimize ADT's new practice of messenger military drill, conducted four days a week with a hundred or so of the downtown ADT boys: "The morning inspection is quite thorough and interesting, the boys going through their regular military drill with the ease and unanimity of old soldiers." The exercises were not kept secret—on July 4, 1873, ADT led about 170 of its 300 messenger boys on a parade through the city: "They marched like old soldiers, and executed the various military movements with remarkable precision, attracting much attention and favorable comment from the spectators."[15]

This was not just a publicity stunt; messengers were also put in a military organization for their day-to-day work. In 1875, the "sergeants" among the messenger boys were the ones who chose which boy ran which message next—thus they wielded considerable power over who could receive tips. "To become a sergeant is the one great incentive to the messengers, for the sergeant wears 'sergeant stripes' on his coat sleeves, and has fourteen buttons on his coat. The 'regular' has eight buttons on his coat and no stripes." Sergeants also earned 50¢ more per week. Western Union adopted the same practice, up until at least the late 1920s: "Messenger captains, lieutenants and sergeants—promoted for merit—make frequent inspections and no messenger who fails to pass muster is permitted to go out with messages." But by this time, the "Western Union Messengers' Brigade" even employed former U.S. Army sergeants to train, two nights a week, any "worthy privates in the messenger ranks," who presumably would then be "rapidly promoted to the rank of sergeant and lieutenant, and then to positions in the clerical or operating forces."[16]

Whether this activity was seen by the boys themselves as unreasonable or enjoyable is open to question. Certainly when the boys were paraded through the city, the atmosphere was a holiday one. That first July 4 in 1873 ended with a picnic, where races were held with prizes of a ball and bat, a ham, and three and five days' paid vacation. However, it is unlikely that the boys were paid for the time spent in drill practice.[17]

No matter what the boys themselves thought, contemporaries would have looked favorably upon ADT's efforts. Both the Civil War and the strikes of 1870 had bolstered support for local militias and military training. According to one historian, in the 1870s "Classrooms were scenes of military-like drill and were staffed by teachers who commonly attributed 'intellectual failure' to 'moral laxity.'" But inculcating military precision really meant exerting temporal control over the messengers. Similarly, creating activities to occupy an employee's leisure time could be considered a form of temporal labor discipline. Military activities were ambiguous in both their intent and effects—managers at some level truly believed they were helping the boys, and boys must have enjoyed the activities at least occasionally. But in the end, they imposed a new "time discipline" on the messengers, forever altering the occupation.[18]

Telegraph operations were arranged not only for time discipline, but for space discipline as well. Early telegraph offices, like the 1846 Pittsburgh office of Henry O'Reilly's Atlantic and Ohio company, were meticulously divided into secure spaces for operators (behind glass walls) and open public spaces for patrons, the press, curious visitors, and messengers. The messengers in particular were expected to manage this public space, to "aid in seeing that the rules are observed, in the preservation of good order in the furniture and writing apparatus of the several apartments devoted to ladies, editors, business-men and travelers visiting the establishment."[19]

As telegraph offices multiplied, messengers began to require new spaces of their own. The old system of simply setting a bench to one side of the room for boys to sit on as they waited for calls—even a regimented and numbered bench—was impractical for many reasons; after all, messengers now needed places to change into their uniforms and to store their street clothes. But the main reason they were moved out of the front office was that with more messengers, there was less willingness to expose them to the customers. The scene of several boys loafing on a bench, arguing loudly, playing craps, or reading cheap literature while waiting for the next call hardly inspired patrons with the proper sense of the modernity and speed of the lightning wires.

Physical spaces of work say much about labor's value and labor's power in the workplace. Part of the social construction of messenger work was that while the boy was meant to be highly visible outside of the office, he was meant to be invisible inside of the office. Outside, alone on a delivery run, the uniformed messenger served as both visual advertising and as the direct customer contact for the telegraph company. Boys were to appear neat, speedy, polite, and responsible, with "Clean Hands and Face," "Uniform Pressed and Spotless," and "Cap Squarely on Head," according to one company diagram. But inside the office, where large groups of messengers waited by necessity since the load-leveling was never perfect (often trying to sleep, due to long hours or odd shifts), they were relegated to back rooms, basements, or upper floors, out of the sight of customers.[20]

These changes began to occur soon after ADT was started. In 1872, WU decided it had outgrown its old headquarters and began to work on a new $1.3 million ten-story skyscraper at 195 Broadway in New York City. In 1875, the building opened, and WU would remain there for the next fifty-five years. The nerve center was the operator room on top, with space for 290 operators, linked by pneumatic tubes to the ground-floor public office. Some 610 of WU's 1,000 New York City employees would occupy the building, handling some 27,000 messages every day. But what of the messenger space in the building? The *Telegrapher* remarked:

In arranging the plans for the new building the messengers seem to have been overlooked, and the space devoted to them is not so well adapted to their use as it should have been. However, it is being fitted up as conveniently as possible, and is a decided improvement upon the quarters previously occupied in the old building. It is above ground, whereas before, the delivery clerks and boys were relegated to the cellar, which was dark and ill ventilated, and almost any change could not fail to be an improvement. The entrance to the messengers' room is on Dey street, entirely distinct from any other entrance, so that the messengers are not brought in contact, and do not interfere in any way with any other portion of the business or building.

Thus in the new building, as in the old, the messengers were segregated away so as not to "interfere" with other telegraph operations—or with the first impressions of telegraph customers.[21]

This pattern was not just confined to New York City, as the 1870s witnessed office remodelings across the United States. In 1876, the San Francisco WU office segregated messengers in the rear of the building, in their own space, away from the main reception area, even passing their deliveries from the clerk through a "small window"; in 1877, the Philadelphia WU office boasted a "partition which separates the employe[e]s from the public" and a separate waiting room for messenger boys, including a private street entrance. By the 1910s, Western Union executives were ordering the remodeling of offices once again, including "the removal of all unsightly metal cages and bars that long had separated telegraph employees from the public." But those employees did not include the messengers.[22]

It is unclear whether the boys were better off hidden in their own space or out in the open among the rest of the office activity. A 1900 report on night messengers described:

A long bench crowded with messenger boys in a stuffy little office on an oppressively hot night. The greater part of the floor space taken by tables containing the rattling telegraph instruments, each table's surface quartered by a low glass partition. In front the coughing of a pneumatic tube and a counter at which the clerks are at work. The clicking of the instruments, the noise of bells and buzzers and the general din inside and out make the waiting-room of the night messenger boys far from restful.

So for some boys, waiting out of sight might have meant better space for rest and entertainment. Either way, changes to office spaces were made primarily for customer impressions, not for messenger comfort.[23]

Messenger spaces shifted not only within buildings, but between buildings as well. In the 1920s, WU centralized many of its New York City messenger operations, keeping only two or three hiring offices throughout the city. Again, this was a form of spatial distancing from the customers, so the constant flow of desperate messenger applicants wouldn't crowd the WU headquarters like a soup line. A striking description of such an office survives in the writings of author Henry Miller, who in the early 1920s, before embarking on his first novel, worked as a Western Union messenger hiring manager. Miller's office, "a low ramshackle building in the downtown section of the city," had three floors. The top floor was a wardrobe depot, from which issued "the pungent, acrid odor of camphor and Lysol"; the second floor contained a tailor shop "where the discarded uniforms of the messengers were renovated, cleaned, and pressed"; and the ground floor consisted of an office in the front and a dressing room in back. The dressing room contained "tiny cubicles . . . partitioned off so as to permit the newly appointed messengers to dress and undress" and "a table covered with sheet metal on which was fastened a huge roll of wrapping paper and a ball of twine" where "he was obliged to wrap his citizen clothes into a neat bundle and make his departure through the rear exit." Miller wrote fondly of the way the back room supported his attempts to have sex with the occasional female messenger applicant: "Usually it was only necessary to throw a feed into them in order to bring them back to the office at night and lay them out on the zinc-covered table in the dressing room." Such managerial transgressions illustrate that hiding the messengers from the public eye could have serious consequences for a workforce with so little power.[24]

How did messenger spaces affect the messengers? Inside buildings, messengers suffered the same inconveniences of noise and dirt, and the same dangers of fire and elevator mishap, that adults did. But as in other industrial child-labor situations, a child's playfulness and lack of experience increased his risk. In April 1875, an Associated Press messenger boy was killed in the new WU building, decapitated by an elevator when he stuck his head into the shaft. Officials mused, "It has been the intention of the company to have a wire screen across the doors, as the boys are in the habit of sounding the annunciator to fool the engineer and see the car run up." And in 1882, a lawsuit was filed against ADT Philadelphia by the parents of a messenger, Willie Burke, who was burned to death while working in the basement of the office, cleaning and refilling the gasoline lamps. One of the lamps exploded, and the boy died within twenty-four hours.[25]

While such events may have been rare, they illustrate that in spite of all the custom-designed spaces for messengers, the difficult nature of the job often overcame any amenity. One reformer describing night messengers in Alabama noted, "The boys who work at night have no place to lie down and may be found on the tables, under benches, or wherever they can snatch a wink of sleep while waiting to be called." No separate room or numbered bench could prevent against fatigue, one of the gravest dangers to the messengers once they left the building. In a job designed for rapid (and often reckless) transportation through the city streets, indoor messenger spaces of rest were intimately related to outdoor messenger spaces of safety.[26]

The movement of messengers around the city was always supposed to proceed with the greatest possible speed, but over the century that messenger boys worked in the telegraph industry, the notion of "how fast is fast?" was constantly in flux as different urban transportation technologies emerged. Historians often delineate three periods of urban transport: up to 1880, through the development of ADT, business was conducted within the compact "walking city"; from 1880 to 1920, the turn-of-the-century "networked city" developed through telegraph wires, water pipes, and railroad tracks, linking broader reaches of space together for a hub-and-spoke daily commute; and from 1920 onward, the "automobile city" developed as personal transport coupled with extensive roads allowed for more decentralized urban and suburban development. This periodization illustrates the options open to messengers for transport in different times; however, walking messengers did not disappear with the coming of the rails and subways, and bicycle messengers did not disappear with the coming of the automobile. For one thing, these different technologies coexisted over a long period of time. And the choice of messenger transportation technology was based not only on pure speed, but also on social factors of age and gender.[27]

In the "walking city," messenger speed first became an issue when ADT began to employ boys to deliver printed intracity messages independent of the "lightning wires." Since the customer cost for an ADT messenger was 30¢ per hour, customers and managers were supposed to compute the expected time of message transit around the city using the various rates between different places as listed in the ADT directory. For example, a boy was allowed twice the number of minutes as he earned for the company in cents, so a 20¢ message would be allowed forty minutes for delivery. By the 1880s, ADT rules stated that "messengers on foot may not take more than one-and-one-half minutes per block" and that boys had to be in and out of destination buildings in under two minutes. This earned ADT boys the derisive nickname "All Day Trotters."[28]

Messengers were monitored and disciplined according to their outdoor speed, so determining a "normal" speed was crucial. Managers recorded the time when a message was given to a messenger, and messengers were to note the time it was signed for by the customer, so "at the close of the day, the time occupied in delivery [could be] averaged and his fitness for service determined." Contemporary news reports indicate what a "slow" delivery entailed: in 1880, the *New York Times* editorialized against incidents where packages took hours to be delivered over distances of only ten blocks, a delivery that was only supposed to take twenty minutes.[29]

As discussed in chapter 3, messenger speed of service was used as a way for telegraph companies to compete without engaging in a "ruinous rate war." Such speed contests were even more important to the messengers themselves, especially when they worked on a piece rate. Two competing district companies would both install free call-boxes for business customers, so those customers learned to buzz both companies and give the message to whoever arrived first (though sometimes a tip would be given as a consolation prize to the losing boy).

With the rise of the "networked" city of faster, more reliable streetcars, messengers had an opportunity to stop "trotting"—but the telegraph companies first had to approve. An 1881 article on a thirteen-year-old ADT boy who was arrested in New York City for stealing a ride on a streetcar illustrates just how the messengers were expected to use public transportation. The messenger explained that even though the company gave him 10¢ for the fare, he was forced to split the money with the office clerk, meaning he could only ride one way; thus he was stealing the ride back. Even when fully paid for, streetcars sometimes didn't make much difference, as one 1886 account shows: "the few car lines we had to depend on were very irregular in the schedule, often causing great delay in the delivery of messages. In the 'Bronx,' where car routes were few, a messenger frequently had to walk from one to three miles on a delivery. The lighting of streets in that borough was very poor, and in some cases the boy had to depend on the light from the houses, which also was a great handicap."[30]

The situation hadn't improved much by 1908. In Chicago, messengers could get tickets from the telegraph office to pay for 5¢ streetcar fares; but these tickets were often traded like cash or even gambled away. Perhaps because of these problems, by 1910 ADT New York City paid streetcar fare for their messengers in a lump sum every year, based on the number of Western Union messages carried, so boys wearing uniforms could ride streetcars for free. This is one reason why ADT didn't want boys wearing uniforms to or from work—they'd get free rides. But this citywide arrangement didn't last. In 1920, Henry Miller sometimes took the car fare that was supposed to go to messengers and used it for lunch money, "doling it out only to 'repeaters' who knew of its existence and were cheeky enough to demand it." Ironically, Miller himself was paid more than ten times what the messengers made—$240 a month—plus a free mass transit pass.[31]

Hidden within the change to a "networked" city was a different shift in transport, midway between the "walking" city and the "automobile" city. In the 1890s, the safety bicycle emerged as a new form of personal transport—sturdy and relatively reliable, expensive at first but fast and versatile. The bicycle as a mode of transportation has been all but ignored by historians in the United States, with the enthusiasm for the safety bicycle in the 1890s described as a "craze" that faded as soon as automobiles became widely available. But this interpretation of the bicycle considers only adult transportation needs and preferences, not those of children. Bicycles remained the main form of transport for boy messengers (not to mention newsies, errand boys, and delivery workers) for another fifty years.

As early as 1868, with the nation's "Velocipede Craze," there were suggestions that messengers could take advantage of the self-propelled, wheeled "boneshakers" to speed their delivery tasks. But not until J. K. Starley's 1885 introduction of the Rover "safety bicycle" in England would such a vehicle be viable for messenger work. Here was a chain-driven bicycle with both wheels relatively the same size, lessening the danger of taking a "header" off of the front of the vehicle (common on high-wheeled "ordinary" bikes). Three years later, Alexander Pope introduced his Columbia safety bicycle to the United States, incorporating solid rubber tires and front suspension. With John B. Dunlop's 1889 innovation of pneumatic rubber tires, the bicycle finally moved from an "extreme sport" to a useful mode of personal transportation, and the bicycle boom began. But even though dozens of different safety bikes were available in 1891, the lowest prices were around $135—still only for the affluent, as this was six months of an average worker's wages.[32]

Soon bicycles had proven their usefulness and dropped enough in price to be taken up by urban service industries. Bicycle police appeared in major cities such as Philadelphia, Cincinnati, and Chicago, and in Brooklyn (where they were known as "fly cops"). The Chicago post office tested the bicycle against other forms of transport, including horse-and-buggy systems and foot messenger delivery, for special mail delivery. They chose the bicycle and outfitted 115 postmen with them in 1895, reportedly saving $5,000 a year in transport costs. And by 1895, the New York and New Jersey Telephone Company (a large part of its stock owned by Western Union) had begun to supply bicycles to linemen and inspectors. Bicycle ads even began to appear in telegraph journals, apparently targeting operators.[33]

It is unclear where and when messengers first mounted bicycles. In 1897, the WU manager at Syracuse, New York, claimed that his office was the first to put the boys on bikes, dating from 1892. By 1894, Philadelphia ADT messengers were using bicycles (and two years later, load-carrying tricycles). But the first detailed report of a bike messenger force came not from the East Coast, but from Omaha, Nebraska, when *Telegraph Age* ran a photo of the twenty-four boys in the "model office" there, fifteen of whom owned $90 bicycles. The journal praised the Omaha managers, not only for having put their boys on bikes since 1893, but also for adjusting the messenger uniforms for better performance on the machines, with a $2 sweater, $3.25 custom bicycle pants, 50¢ belt, and 50¢ badge, "a fine advertisement for the telegraph company."[34]

Soon afterward, ADT New York City messengers were using bicycles too: "In the majority of cases the riders own their wheels, but the telegraph company has furnished a few and expects to purchase more." However, ADT would not put boys in the "crowded lower part of the city" on bicycles. The bikes were portrayed as a staff-cutting tactic, like any good machine: "While it does not increase the company's expenses to any great extent, it decreases the number of messengers and increases the revenue of those employed, as the company pays so much [per] message for delivery." The ADT managers reported that the payback came only on long runs, because on short runs the time needed to get on and off the bicycle (and to find a safe place to park) ate up the time saved by riding.[35]

By 1902, after the entry of department stores into the bicycle-selling business, a spate of overproduction, and the collapse of the feared American Bicycle Company "bicycle trust," a new bicycle might cost only $15. Here was a price the messengers could afford, and at this point most telegraph offices tried to hire boys who already owned bikes. By 1904, the *Commercial Telegraphers' Journal* could run a cartoon showing an "evolution"-type progression of messenger boys and their transportation options: 1890, walking with cigarette; 1900, on bicycle; 1910, in automobile; 1920, in airplane (Wright-brothers style); and then, as an angel with a harp (implying that these transportation modes were progressively more dangerous) (see figure 4.2). Plus, the journal was running ads for bicycles, now targeted to messengers. Sears claimed, "$11.75 BUYS OUR NEW 1906 MODEL KENWOOD, the long famous $75.00 bicycle, strictly high grade, worth three of the bicycles sold by others at $15.00 to $25.00."[36]

By the 1920s, bicycles were so necessary to messenger work that WU bought them from the Westfield Manufacturing Company (makers of Columbia, Westfield, Tribune, Rambler, Crescent, and Pope brands), and resold them to their messengers at cost. In 1938, the Westfield sales manager reminded WU's messenger equipment manager, "Western Union contract prices are available to you if at any time you want bicycles for any of your particular friends." In fact, during World War II, Westfield was the only company permitted to continue producing bicycles under the War Powers Act. Messengers repurchased these bikes from WU on installment, resulting in another fee taken out of their salary in addition to uniform fees. One former messenger remembered getting his bicycle in 1934 in Marietta, Georgia. He purchased the bike from WU on the installment plan for around $22: "I rode the street car to Atlanta 20 miles away, assembled it in the basement of the

Figure 4.2. *"The Rise of the Messenger Boy,"* 1904

Source: *CTJ* (July 1904), 6.

Western Union Bldg. 56 Marietta St. Atlanta, and rode it home to Marietta (highway 41) to save express charges."[37]

Like the uniforms, bicycles represented serious expenditures for WU. Compare the money WU spent on bikes for their messengers to the money they spent on shoes for the boys (see figure 4.3). From 1923 to 1942, WU purchased bikes in bulk, sometimes in orders up to five thousand per year in 1929 and 1930. These bikes cost between $20 and $25 each, and were resold to messengers. WU also purchased and resold bulk bicycle supplies, mainly tires. Similarly, from 1925 to 1939, WU purchased pairs of shoes for messengers in bulk, sometimes in orders up to fifteen thousand and sixteen thousand in the peak years of 1931 and 1936, during the Depression. Pairs of shoes cost between $2.75 and $3.50 each, and were resold to messengers. But outfitting the walking messengers was much cheaper than outfitting the bicycle messengers—it took $120,000 to purchase 5,000 bicycles, but that same $120,000 could purchase a staggering 32,000 pairs of shoes.[38]

Selling the bicycle to the messenger along with the uniform was not a coincidence. Both accessories were intended to enhance the efficiency of the messenger while serving as new measures of control. Even the 1869 call for messengers on velocipedes painted the machines as not only vehicles of speed, but vehicles of discipline: "It has all the elements of excitement which would suit and captivate a spirited boy—a boy of pluck and metal [*sic*]. There would be no windows for him to gaze at—the gaze would be at him. He would glory in showing to admiring crowds how he could spin along on his mission, blowing his whistle at the crossings, and would claim 'good time' on his return" (see figure 4.4). By 1898, after the first successful attempts at messenger bicycle use, the *Telegraph Age* agreed that "better-class lads" were attracted to the messenger service because the bicycle encouraged the boys to compete at speed and not

Figure 4.3. *Western Union bicycle and shoe expenses, 1923–1942*

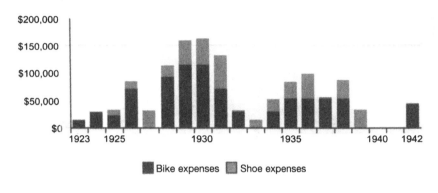

Source: WU, executive minutes.
Note: Blank years may indicate missing data.

at craps. Ironically, a decade and a half later, one Boston reformer would argue that because the vehicle had become such a necessity for the job, "the messenger boy's greatest weakness is stealing bicycles."[39]

With all the interest in bicycles, did messengers ever participate in the "automobile city" at all? As early as 1939, WU began to resell motorbikes called "Telemotors" to messengers in addition to bicycles, and by 1941 so-called motor messengers using both Telemotors and automobiles had taken over all routes over ten miles. But because of driving license restrictions, these were always older teen or adult messengers. In some ways, their situation was quite similar to that of the bicycle messengers, in that they usually had to provide their own vehicles and pay for their own repairs. But motor messengers, because of their age and because of the adult associations with cars, were never considered "children" like bicycle and foot messengers. They earned higher wages and were even covered under union contracts as adults.[40]

The irony of restricting the bicycles to the children is that bicycles really did make sense for the messenger job—they were adopted in the 1890s not because they were inexpensive, but because their cost was justified by their speed; not only because they were an enticement to the employment of male children, but in spite of the fact that most bicyclists were adults (and a substantial portion were women). However, when the bicycle boom ended in the early 1900s, the automobile came to be seen in the United States as the adult's vehicle, and the bicycle as the child's. Even though there was a mini-revival of the bicycle with the coming of the Great Depression, bicycle shops catered mostly to kids, selling $40 bikes for "$5 down, $5 a month." By 1941, fully 85 percent of all bicycles produced were children's vehicles.[41]

Figure 4.4. *Cartoon praising the new bike messenger, 1895*

THE OLD WAY OF DELIVER-ING TELEGRAMS.   THE NEW WAY OF DELIVER-ING TELEGRAMS.

Source: *Telegraph Age* (1895), 466.

Thus the bicycle, which for functional reasons was an excellent machine to use for individualized, local small-parcel and message delivery, served at the same time to help maintain the idea that a messenger's job was a child's job—and it was consciously used by WU to attract new messenger boys to the business. One former messenger recalled his 1938 job in Dover, New Hampshire, with fondness, saying the best part of his work was "Getting paid to ride [a] bike."[42]

But in a very important sense, young bicycle and foot messengers were a part of the "automobile city" even though they themselves were not in autos. After all, they shared the same streets with cars, and were now exposed to a host of new risks. Messengers were all alone in an environment increasingly populated by more and more traffic conveyances—from horses and carriages and street railways of all types, to motorcycles and cars and trucks. Traffic laws trailed behind such advances, and with the piece-wage imperative of delivery speed, safety was bound to suffer.

This was not just an urban problem. The fact that the telegraph network followed the railway network meant that rural telegraph offices faced the dangers of the rails. One telegraph messenger from Opelika, Alabama, died in a railroad accident outside of his telegraph office: "Our messenger boy, Master Ernest Doughty, was walking on the railroad track, and the switching engine ran over him, severing his head from his body." Such dangers faced operators and lineworkers as well—telegraph journals routinely advertised artificial limbs to their readers. But it was the messengers who were coming and going in all directions as the day went on.[43]

For messengers, the big cities were the sites of the greatest danger to the greatest number. Messengers in the walking city faced down horse-driven hacks, but trains were the greatest danger in the networked city. Historian Viviana Zelizer argued that "the first two alarming newspaper stories on child death" in New York City streets appeared in 1904, concerning children hit by streetcars; however, most of the blame was directed toward parents, not toward the transportation companies or toward any businesses that may have been employing the children at the time they were hit. Concern over such accidents grew until in November 1908 five hundred New York City children marched down "Death Avenue" (as Eleventh Avenue was known) to demand that the New York Central Railroad remove its train tracks from their neighborhood. By 1913, when the Brooklyn Transit Company organized its first "Children's Safety Crusade," lecturing in public schools and distributing safety literature, between 40 percent and 60 percent of New York City traffic victims were under age fifteen. One reformer even referred to the epidemic as "trolleyitis."[44]

Yet only after the coming of the automobile in the 1920s did the telegraph companies apparently think about safety. For example, Henry Miller wrote of his 1920s office, "The walls were plastered with safety posters, photographs, and written sheets giving the latest country-wide statistics relating to messengers killed, crippled, or incapacitated." The telegraph industry warned young messengers about street hazards, but did not remove them from those hazards. A 1922 book for telegraph messengers, prepared with the cooperation of WU, included a series of lessons on "Safety, Health and Hygiene." These included "The Human Walking Mechanism," "Care of the Feet," "Safety First!" "Correct Posture for Cyclists," and "How to Walk." A few years later, PT published a list of "hints to messengers" alongside a tale of a messenger who "fell from his bicycle" and had to be hospitalized for several days, advising, "WALK across a street—never run. A sudden dash may mean death." In both cases, the assumption was that messengers had to be educated on walking and biking in order to avoid fatigue and injury, not that the work pattern had to be changed or slowed down.[45]

No doubt injuries occurred regularly—in 1922, 477 children died in street accidents in New York City alone—but stories of messenger accidents only appeared in the telegraph press when messenger heroics were involved. In 1921 when a San Diego messenger—"Little Hector Toreres" of "Spanish origin"—was hit by a car while delivering a message, it was reported that while laying in his hospital room with "painful but not serious" injuries, he could think only of his undelivered message. Similarly, a 1924 article based on an extensive interview with the New York City WU messenger manager claimed paradoxically, "The messenger service is not hazardous; but it is an exacting service. It often calls for resourcefulness and outright courage."[46]

Just as with other aspects of the telegraph network, safety varied with geography. As described earlier, Western Union did not allow bicycle messengers in lower Manhattan, ostensibly because of safety reasons but probably more because walking was actually faster, considering the dense coverage of branch offices and the time spent inside buildings on elevators or stairs. Of course, even for foot messengers, any directives to "walk not run" were in direct conflict with the messenger's expected duty of speed, and the messenger's own need under the piece wage to finish each run as quickly as possible in order to fit a large number of runs into each shift.[47]

Small cities and rural areas had their own concerns, not only because bikes were more common, but because such areas were ill equipped to handle bad weather. In 1935, there was an ice storm in Marietta, Georgia, and a former messenger there described how that week it was "Almost impossible to even stand on the ice, much less walk or ride a bike. We wrapped our shoes in burlap, drove roofing nails through our shoes, tried baseball shoes with metal cleats, anything." Hills were the worst: "I crawled at least one fourth of a mile up a hill one night to deliver a telegram to a Mr. Glover. I was on my hands and knees, crawling zig-zag from one bush to another or to a tree or rock or stump or anything to keep from sliding back down the hill and to push up the hill a bit—but I delivered the telegram and as best I remember—NO tip."[48]

Whether due to the elements, to poor vision at night, to job-related haste and fatigue, to ill-maintained equipment, or to the sheer volume of traffic on the urban streets, messenger accidents did occur. Besides costing the messenger in time away from piece work, accidents could have real costs to the telegraph companies—not only in the occasional doctor bills they paid, but also in lawsuits when speeding messengers hit pedestrians. On November 25, 1924, WU paid $12,000 to settle a personal injury suit brought on behalf of Thomas Malarkey, who died in Marion, Ohio, when a bicycle messenger collided with him. A year later, WU paid $3,000 to settle the claim of Eva P. Albritton, a Jacksonville, Florida woman who sustained injuries after being hit and knocked down by another WU bicycle messenger. Such incidents compelled WU to purchase insurance covering its automobile and bicycle endeavors early in 1926. But even if bystanders were compensated, messengers were not always reimbursed for injury or damages. One former messenger described how, while making a delivery, he once laid his bicycle on the ground in front of an office only to have a car back over it. Although WU demanded that messengers purchase bikes as a condition of employment, WU would not reimburse him for the cost of a new bike (roughly $32). Fortunately, the person who damaged it decided to "do a poor kid a favor" and paid for a new bike.[49]

Messenger safety didn't get much press until 1932, when the White House Conference on Child Health and Protection requested data on work accident rates of minors related to vehicles. The Children's Bureau subsequently collected data on telegraph messenger accident rates for 1931. In 1934, the report appeared in the *Monthly Labor Review*, indicating that from 1924 to 1932 the messenger "accident rate" fluctuated between 13 and 20 percent. Of the messengers polled, 30 percent were foot messengers, 66 percent used bicycles, and 4 percent used motorcycles or autos.[50]

Foot messengers suffered proportionately the fewest injuries and the least serious injuries; motorcycle messengers, proportionately the most and the worst. Automobiles and trucks were the greatest single causes of accidents, especially to bike messengers. And 21 percent of the injuries occurred at night, whether due to poor vision, fatigue, or both. Even though both WU and PT paid accident disability benefits to messengers, the authors pointed out, "The benefits paid reflect of course the low wages paid immature workers in an occupation requiring no skill, special training, or previous experience."

The report had mixed results. Union representative Mary L. Cook attacked WU for "underpaid messengers being kept from work because they could not afford to pay repair bills for their dilapidated bicycles." Western Union responded by promoting the "regular courtesy and accident prevention meeting" assemblies that their messengers attended, and the motivational safety materials they placed in offices, such as signs saying, "Our Office Has Gone ____ Days Without An Accident. Help Add Another Day!" WU touted the effectiveness of such training when it awarded to its Philadelphia messengers a national safety award, given because in 1937 the 350 messengers had "only 0.29 accidents per ten thousand hours worked." This benign-sounding statistic, when computed over 350 full-time messengers, works out to about twenty-five accidents that year—or one accident every two weeks in the city. Whether contemporaries would have considered this a "low" number or not, when an urban telegraph network is viewed as a whole—even this award-winning network in Philadelphia—it becomes clear that messenger accidents were not isolated events, but an ongoing cost of doing business.[51]

The safety issue kept returning in the many government hearings on the telegraph that ran through the late 1930s and early 1940s, in anticipation of WU's merger with PT. In 1938, a night bicycle messenger from Brooklyn testified in a letter to the Senate Committee on Interstate Commerce, "Bicycle messengers, who are usually employed in the residential sections of the city, are by far the most exploited workers in the Western Union." He told the committee that they worked up to sixty hours a week and may earn as little as $6 or $10 per week for that effort. Out of this money they had to pay for bicycles, equipment, and maintenance. As for safety, "Protests by messengers that it is dangerous to ride are often ignored by company executives." He said messengers were forced to ride in snow and ice, late at night, or when they were physically exhausted.

And he cited the case of messenger Walter Ebinger, killed by a hit-and-run motorist in Brooklyn at 2:00 A.M., saying this bicycle messenger was forced to work on his one day off and was "physically exhausted and half asleep."[52]

The safety question persisted well into the 1940s, nearly to the end of the boy messenger service itself. In 1945, WU assistant vice president T. B. Gittings testified before the Senate Committee on Education and Labor, saying, "We have a legal as well as a moral obligation to the messenger and his parents to exercise great care for the safety of the individual." Gittings argued that WU had opened more offices in downtown areas so that messengers could go on foot instead of on bicycle (highlighting the fact that walking was safer). WU had also been replacing bike messengers with autos (driven by those eighteen and over), and had abandoned using motorcycles. "Safety principles are constantly drilled into messengers," he said, and "safety incentive contests" and "mock court sessions" had been tried. Bicycles were allegedly inspected daily. And of course, Gittings reminded his audience, "hazards exist in going to and from school."[53]

Hazards certainly did exist in going to and from school, but then again, children going to and from school weren't racing to a random area of the city in order to pick up a message before a competing boy did, and then racing back again in order to cram the maximum amount of 2¢ calls into an evening. The speed demands of the telegraph industry, both from the manager's point of view and from the messenger's, could result in the innovative use of technologies such as streetcars, bicycles, and autos; but those same demands could result in fatigue and recklessness, especially since messengers outside the building were accountable to no one but themselves. Each of the transportation methods available to the messenger embodied within it a certain form of temporal, spatial, and social discipline that could not be ignored.

Taken together, the messenger innovations made by ADT and WU—drills and numbers, uniforms and bicycles, military and mechanical metaphors—altered the messenger boy occupation like never before. Messengers were no longer in an apprentice relationship with kindly operators, but in a temporary employment relationship with a corporate middleman. Yet messengers' movements through time and space, enabled by their accessories of uniforms and bikes, still set the speed at which telegrams traveled in the city. The next chapter will consider this relationship between message and messenger more closely.

CHAPTER FIVE

THE MESSAGE AND THE MESSENGER

A change comes over a boy the moment his first message is placed in his hand. Intuitively he realizes, at that instant, that he is intrusted with a genuine and grave responsibility. He seems to feel a touch of the subtle magnetic current—the soul of telegraphy!—and grasps the vital, energizing realization that time is the essence of things.

—*Telegraph Age* editorial, 1902[1]

In the ninteenth century, telegrams were put to both business and social uses, but the relationship between telegram cost, length, and distance meant that businesses were more likely to use the telegraph than were individuals. Brokers could be relied upon to regularly send large amounts of expensive, wordy, distant telegrams, and so the entire temporal and spatial telegraph network shifted to accommodate their needs (as described in chapter 2). By 1887, Western Union president William Orton estimated that only 2 percent of the public ever sent a telegram, and that only 5 percent of all telegrams sent were personal communications.[2]

But the nature of telegraphic communication had changed by the turn of the century. First of all, for the most lucrative telegraphic services, special technologies and special departments had emerged to serve business needs with new commodities besides telegrams. (In fact, the telephone could be seen as just this kind of spinoff from the telegraph.) Second, as businesses moved away from using telegrams, especially due to competition from the telephone, telegraph companies were compelled to redefine the purpose of the telegram and reeducate businesses in the best ways to use telegrams in concert with other media, such as in national sales promotions. And third, both the decline in business traffic and the spatial/temporal unevenness of telegraph service led managers to seek more load-leveling strategies by appealing to nonbusiness telegram consumers, whose greeting-card traffic could be used to keep the telegraph network busy during the lax times when business demand dropped daily, weekly, or yearly. But these three strategies had contradictory results, simultaneously defining telegrams as serious business and as playful greetings—conflicting commodities that served to further complicate the messenger boy's role as well.

Telegrams initially carried all sorts of business information—market quotes, buy and sell orders, and commercial news in particular. But businesses soon learned to economize in their use of telegrams, creating vast code books of telegram shorthand, enforcing strict rules of telegram composition, and generally putting more time, effort, and money into crafting cheap telegrams themselves instead of paying the telegraph company for a wordy transmission. Such moves provided an ongoing incentive for the telegraph companies to develop new communications commodities that catered specifically to business needs.

Commercial news was the first category of telegraphic information to spin off. When the United States went to war with Mexico over Texas in 1846, the telegraph had its first real test as a news-delivery service. Soon most newspapers offered "telegraphic dispatch" sections, featuring financial and commercial information, plus occasional reports on governmental debates or natural disasters. This function fit in well with load-leveling demands, as long news reports at special reduced rates were transmitted during slow times of the day. In this case, the telegraph was used not for point-to-point communication, but for sending a single message simultaneously to many different offices at once, known as making "drop copies." For example, press dispatches sent twice a day from New York City to Buffalo (five hundred miles away) also showed up at intermediate places such as Albany, Utica, Syracuse, Rochester. The largest of the news services, the New York Associated Press, did over $50,000 of business with the telegraph companies in 1852.[3]

Following soon after the spinoff of the news telegrams were those telegrams dealing with market information. As early as 1865, a separate "gold room" was established in the New York Stock Exchange, with messengers running the latest price quotes back and forth to brokers. The potential for mechanized delivery of this information seemed clear, and soon a telegraph-operated, numbered disk indicator was developed and installed in the offices of fifty subscribing brokers. By 1867, three hundred brokers subscribed to this new system, now incorporated as the Gold Indicator Company.[4]

Enter Edward Calahan, again. Before he invented the messenger call-box, Calahan developed the first stock ticker, which continuously printed price quotes on paper tape. On the basis of this new device, a new firm, Gold and Stock Telegraph Co., was organized in 1867 and consolidated with the existing Gold Indicator Co. under Western Union. Unlike the call-box, the stock ticker was meant to *reduce* the need for messenger boys; however, whenever the ticker system went down, each of the company's three hundred subscribers sent a messenger boy down to the exchange to find out what had happened.[5]

Thomas Edison later worked on improving the gold ticker, and Gold and Stock eventually signed a ninety-nine-year lease with its parent company, bringing its eight hundred subscribers together with the existing Western Union Commercial News Department (already connected with the New York Associated Press). Western Union extended the stock ticker system to other cities through this central department, with nationwide stock quote telegrams known as "CNDs" sent out to subscribers on a regular basis. Although some quotes were handled over regular Morse lines through human operators, most of this traffic flowed over special wires and to special receiving equipment. As a bonus, the CND business generated a return flow of messages to buy or sell stocks and commodities: another special type of short, high-priority telegram called a "market order" or an "OD."[6]

In this way, the more telegrams were used for particular purposes, the more likely it was that the telegraph companies would create spinoff products capturing those particular purposes with particular technologies. Even the well-known practice of calling up the telegraph office to find out the current time led WU to offer its "telegraphic time service," where wired signals ran automatic clocks for a fee. Spinoff services still generated revenue for the telegraph companies, perhaps even at a higher profit since the recommodification of the information was often more mechnized and involved less labor. But at the same time, each spinoff service made the plain old telegram less and less useful to business.

The last spinoff was perhaps the most damaging to telegram sales, because it was the most versatile spinoff of them all: the practice of allowing customers to lease entire telegraph wires themselves, so they could move their own information entirely outside the labor structure of the telegraph company. As described in chapter 3, large businesses had begun to set up private telegraph lines using Morse equipment and employing skilled Morse operators as early as 1850. The telegraph companies eventually decided that this was a market they needed to capture themselves. Leased-line agreements were pioneered as early as 1867, but it wasn't until 1884 that WU began to regularly lease wires to private brokers, retail houses, and banks. The leasing companies employed their own operators and had exclusive use of a telegraph company wire, but only at certain times of day. This business was apparently so profitable that WU seriously considered abandoning its own handling of messages entirely, and merely leasing wires out to others.[7]

Figure 5.1. *Western Union line rental, 1893–1908*

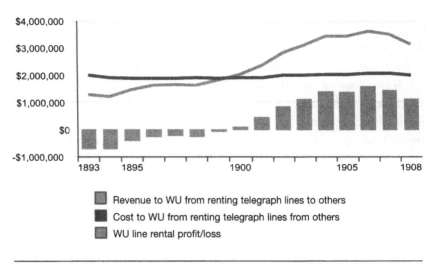

Revenue to WU from renting telegraph lines to others
Cost to WU from renting telegraph lines from others
WU line rental profit/loss

Source: WU, statistical notebooks prepared for Robert C. Clowry (1893–1908) WUA, 1993 addendum, series G, box 81, folder 4.

Wire leasing was attractive because it was yet another way to load-level fluctuating message traffic over a fixed-wire plant and through a fixed labor force. A single wire could be used several different ways during a single day: a broker could lease it from 9:00 A.M. to 3:00 P.M., to send ODs while the market was open; the telegraph company could use it for public service from 3:00 P.M. to 8:00 P.M., sending messages in batch that had been waiting around all day; and then a newspaper could lease the wire from 8:00 P.M. to 9:00 A.M. for overnight transmission of long articles. By the early 1900s, leasing clients paid a yearly fee of $20 per mile of wire; shared use of a wire from New York City to Philadelphia cost $1,800 per year in 1909. This was the most profitable part of WU's business at this time, which makes sense considering it was the service involving the lowest labor costs. In 1900, the revenues that WU earned from leasing out its own wires finally exceeded the costs that WU paid to other telegraph and telephone companies for the right to use *their* wires[8] (see figure 5.1).

From the customer's point of view, however, leased lines still demanded skilled Morse operators until World War I, when reliable page-printing telegraph equipment was finally available. Ironically, this equipment came not from Western Union, but from AT&T. Even though they were a telephone company, AT&T dealt a brisk business in leasing their lines for private telegraph use just like WU did. In 1909, just as it was taking control of WU (described in chapter 7), AT&T commissioned Western Electric to develop a "page printing telegraph system" in order increase their private-line telegraph business. AT&T's first such service was slow and unreliable when it debuted in 1915, but by 1919 the printers operated at sixty words per minute and could transmit and receive reliably over short intercity distances. By the 1920s, "scientific office management" experts advised companies to install "telegraph typewriters" to avoid the charges of the telegraph companies and the unreliability of messengers, especially "in those large organizations where hundreds of messages are received daily, and where the old system necessitated that each of those messages be typewritten before it could be transmitted." And in the 1930s, AT&T's now-mature Teletype network made the telephone company into WU's greatest competitor, with Teletype advertisements proudly claiming, "This *Electric Messenger* never idles—never loses or misdelivers messages—never meets with injury."[9]

While AT&T was pushing ahead with its leased-line telegraph services, WU experienced a management crisis. On the one hand, they wanted to compete with AT&T, and they did develop their own automatic leased-line service. In 1932, WU's "Timed Wire Service" (TWS) used special automatic typewriters in subscribing offices. The customers sent and received messages themselves, and were charged by the telegraph company for the amount of time they monopolized the line. Customer-employed operators could send one hundred words in three minutes, which meant that TWS cost only twice as much as a regular ten-word telegram. But on the other hand, WU wanted to continue the lucrative practice where large companies simply sent books of telegrams back and forth between their units each day. One former WU manager recounted how he was assigned by WU to work six weeks at the General Electric headquarters to analyze their telegraph traffic with the other GE plants. His recommendation was that GE lease a series of twenty-four-hour telegraph lines between its headquarters and its major branch plants, in order to save money and still send more telegraphic traffic. Soon "all hell broke loose" when he reported this arrangement back to WU. "I was told that I was simply throwing the Company's money away, and starting a very dangerous precedent." But he argued that if WU didn't strike such deals, AT&T would.[10]

Did the spinoffs cut into basic telegram service as WU officials feared? In 1908, telegram volume still made up over three-quarters of Western Union's business. About 125 million WU telegrams were sent at the end of World War I in 1918, growing steadily to an all-time high of 200 million in 1929, before the Great Depression. Still, WU believed that businesses were not using as many telegrams as they should—and that AT&T's increasingly popular long-distance phone service was to blame.[11]

Competition with the telephone highlighted the hidden labor involved on the consumer side in crafting an efficient telegram, especially in staying under the ten-word limit of the base telegram rate. One 1919 business textbook recommended typing each telegram out in all capital letters with no punctuation before it was sent, to see exactly how the message would be received and make sure it still made sense without, say, commas and periods. Another text from the same period advised senders to use code books when crafting telegrams—either the proprietary, secret code books created by large corporations, or the public code books issued by business publishers. Such code words were subject to strict rules: they had to be made up of ten letters or fewer, and could consist only of actual words, combinations of words, or "artificial words" that were "pronounceable" in at least one of eight Roman-alphabet languages (English, French, German, Italian, Dutch, Portugese, Spanish, and Latin). For example, in *Lieber's Standard Telegraphic Code*, "autogeneal" meant "may not have such an opportunity again"; "autogony" meant "the opportunity will be lost unless you telegraph quickly." While using such codes to stay under the ten-word cutoff could certainly save on telegram cost, it added significant labor time to the construction of the telegram itself—and to the reconstruction of the telegram's message on the receiving end.[12]

As early as 1869, Western Union explored the possibility of creating a "system of night messages at reduced rates" to remove the need for such complicated tricks. But ironically, the telegraph company introduced cheaper telegrams only after it was taken over by Theodore Vail's telephone company in the 1909 AT&T stock buyout of WU (described in chapter 7). Vail inaugurated "night letters" and "day letters," which offered flat rates over any distance for fifty words or less (no codes allowed). The catch was that night and day letters were delivered at the telegraph company's convenience, not the customer's. Night letters, for example, were accepted up until midnight but not delivered until sometime the next morning, and thus traveled between cities when the wires were least crowded. But there was also another load-leveling strategy at work. Messengers were already employed at night, and night letters increased their evening tasks, since messages received in the wee hours of the morning had to be physically moved from the central office to branch offices before the next day's delivery hours began.[13]

Day letters were similar: they were accepted in the morning with delivery attempted (but not guaranteed) sometime that very same day. The day letter was 75 percent cheaper than a regular telegram, but more expensive than a night letter. Similar restrictions applied to day letters: they could not be coded, they could be deferred for the regular day traffic, and they could be delivered by phone if the telegraph company desired. This last caveat introduced a crucial difference in telegram service, as what were now known as the "black" or "full-rate" telegrams were still expected to be delivered immediately by a human messenger in all cases, but the "red" night letter and "blue" day letters could be both delayed and telephoned.[14]

Because of their restrictions on coded words, their longer delivery times,

Table 5.1. *Western Union telegram revenue by type, 1926–1940*

Year	Revenue from full-rate telegrams	Total revenue (%)	Revenue from reduced-rate letters	Total revenue (%)
1926	$75,317,000	66	$32,279,000	28
1930	$68,401,000	64	$31,142,000	29
1934	$40,399,000	57	$21,850,000	31
1938	$37,792,000	52	$20,515,000	28
1940	$43,706,000	55	$18,910,000	24

Source: McKay (1931); U.S. Senate (1941).

and their allowances for phoned delivery, night and day letters were not as popular with businesses as with individual consumers. Western Union was walking a fine line between creating a new telegraphic commodity to make better use of its physical and labor resources, without at the same time undercutting its expensive and profitable regular telegram service. Over the 1930s, the revenue comprised by the plain old black telegrams began to drop, both in raw numbers and as a percent of total message revenue, but the industry still depended very much on this old standby[15] (see table 5.1).

These numbers don't reveal how many full-rate telegrams were for business purposes, and how many were social greetings. But such telegrams still made up the bulk of WU's revenue, no matter who was purchasing them. The challenge for WU was to make its telegrams more profitable by reducing their labor requirements, and to sell more telegrams by appealing to a broader market. Both strategies implicated the messengers.

With both leased lines and the new night and day messages stealing business away from the full-rate telegram through the 1920s and 1930s, WU found itself reeducating businesses and consumers alike as to the use of their various telegraphic products. One campaign focused on ways to use the messenger's package-delivery abilities. As early as 1874, ADT had set up a "circular service" for handing out advertisements door-to-door. This service grew to encompass the delivery of everything from wedding invitations to political literature, targeted to a specific set of consumers in a specific neighborhood. By 1875, ADT claimed that "200,000 unaddressed circulars can be delivered by this Company, one or more at each house *within twenty-four hours.*"[16]

This was one of the first direct-marketing services ever, growing by the 1880s into a nationally coordinated effort. ADT kept a list of 150,000 "Special Names and addresses of desirable Individuals and Firms in the Principal Cities of the United States" and, with its Western Union links, could send ads simultaneously all over the country. ADT saw itself as competing directly with the post office, claiming, "Our price for furnishing envelopes or wrappers, addressing, folding, filling and delivering is *less* than postage alone."[17]

Again, load-leveling was at work. ADT was proud that such feats were achieved "without an increase in the messenger force usually employed." But the service grew to involve more than just the boys. By 1884, ADT claimed that "A special staff is employed for the purpose of Addressing, Folding, Enveloping, etc., Advertising Matter for Mailing and other methods of distribution." Adult managers or "roundsmen" were also needed to monitor the messengers, so that "any attempt by the messengers to evade their duty, by throwing away the circulars or otherwise, is morally certain to be detected and punished." One story described how during a turn-of-the-century Chicago city council race, messengers were hired to deliver flyers that urged people to vote for one of the candidates. Recipients were outraged, because they initially assumed that the "telegrams" were bearing bad news. Confronted with this situation, some of the messengers gave up and sold the bulk of their flyers to a rag picker and then went to the movies with the proceeds (since they were still on the clock).[18]

Beginning in the 1920s, the circular service was promoted even more aggressively, as an adjunct to nationwide radio and print campaigns—messengers would deliver samples of soap, cereal, and other household products. As "scientific office management" guru William H. Leffingwell commented in 1926, "It has been only in the last 15 years or so that the business world realized that direct-by-mail advertising is one of the most potent of all marketing or mass selling methods." In fact, at the very moment when regular telegraph traffic was plummeting, WU noted with glee that circular business "blossomed" from 1930 to 1933, with the coming of the Depression. Postal Telegraph even claimed, "This has permitted us to employ more messengers than would otherwise be the case."[19]

Western Union called this new advertising strategy "Dramatized Delivery": "Advertising matter or sales presentations that would be tossed into the waste basket if received by mail, get consideration when dignified by ADT delivery." In 1938, "Crowds on a New York street were astonished recently to see forty Western Union messengers with their arms full of cages containing pigeons," a promotional stunt for the magazine *Harper's Bazaar*. As late as 1950, a WU messenger manager wrote, "The uses to which Western Union's special messenger service could be put are almost unlimited. There is scarcely a business in America today which couldn't gain added profits and prestige through a Western Union promotion."[20]

But Dramatized Delivery relied for its prestige on the image of the messenger. The very uniform of the messenger, coupled with the importance of a telegram, made the messenger a unique advertising vehicle, since an individual would stop to receive the ad pitch on the authority of the messenger's presence. In this way, the messenger boy added a speaking voice (and a smiling face) to the so-called silent salesman, a trunk packed full of merchandise and free samples that a manufacturer might otherwise ship to retailers. Plus, the nationwide employment of messengers allowed the telegraph company to simultaneously deliver the same advertising message throughout the country—a spatial and temporal first.[21]

Dramatized Delivery had another payoff for WU as well. When businesses used messenger-delivered telegrams to sell their own products to consumers, those messengers were also implicitly marketing Western Union telegram services to those very same consumers. Selections from a surviving list of "Messenger Distribution Service Proposed New Photos" (complete with cross-outs and corrections) show some of the target markets that WU hoped its advertisers would court:

1. Messenger setting up small grocery display while grocer reads ~~Spl. Sta. msg.~~ *pink*
2. Messenger delivering folder to attractive housewife. (Residence where delivery being made should be modern and in good repair.)
3. Messenger delivering sample to attractive girl at desk in office.
4. Messenger~~s~~ (2) distributing samples at busy street intersection. ~~(Supervisor in background.)~~
5. Messengers (2) distributing samples at factory exit as employes leave. (Supervisor in background.)
6. Messengers (2) distributing samples at theatre as patrons leave. (Supervisor in background.)
7. Messenger placing jumbo telegram on a store window.
8. Messenger with pad and pencil marking down answer to ~~verbal~~ question he has asked attractive housewife.
9. Messenger standing on street with clocking device in his hand.

Whether small businesses, passersby on the street, or "attractive housewives," each of the desired targets of Dramatized Delivery were also potential Western Union telegram customers.[22]

But did businesses use Dramatized Delivery at all? Consider the reaction of Western Union's competitors. During the 1941 Senate investigation of the telegraph industry, the Trade Association of Advertising Distributors, a group of companies that distributed circulars, coupons, and catalogs door-to-door and office-to-office, complained of unfair competition from the telegraph companies. This was a big complaint from an industry that did $125 million in gross business, involved 2,000 companies, and employed 100,000 workers, a large percentage of whom were "adult mail carriers."[23]

The trade association argued, "A substantial part of the messenger service offered by the telegraph companies consists of the distribution of advertising matter and samples, and in that connection these telegraph companies are engaged in the advertising distributing business." Messengers were key to the unfair competition charge:

> The telegraph companies dress their messengers, who are used to distribute advertising matter and samples, in the same distinctive uniforms that are worn by the messengers who pick up and deliver telegrams. These distinctive uniforms have become identified in the public mind with important and urgent telegraph messages, and because of this, advertisers are influenced to engage the telegraph companies to distribute their advertising matter and samples.

The group also protested the way the telegraph companies employed boys at low wages, while their own members employed men at high wages.

But the trade association was mistaken when it argued that "The telegraph companies hire special boys to do distributing as and when needed, and rarely, if ever, use any of the regular messengers." Even when offices hired boys specifically for one or the other duty, the boys would invariably be transferred to wherever the need was greatest, doing whatever job had to be done. Western Union dreamt up countless ways to bring the telegraph to the consumer more directly, and these strategies all involved messengers as sales agents. At first, messengers were merely meant to advertise telegraphic service as an accessory to their own duties (one of the original purposes of uniforming them). Messengers were of course told to solicit replies to messages, but they were often sent out to solicit new business as well, even given bonuses or held to quotas for seasonal campaigns.

Astute managers realized that the messenger's face was often the only face the customer ever saw—the face that the customer associated with the telegraph company as a whole. The manager of the Ft. Worth, Texas Western Union office said in 1915, "My messenger force is one of the best advertisements we have—they are all fine boys and on the job at all times, in full uniform." The manager of the Chicago WU messenger department agreed in 1920, "The messenger boy is a company representative and is often the only one ever seen by the patron whose office is equipped with a call box. To a degree, much greater perhaps than many realize, his deportment, his appearance and his courtesy or lack thereof are potent factors with our patrons in forming an estimate of the company as a whole."[24]

Not only were living uniformed messengers used as advertisements, but illustrations, cardboard cutouts, and mannequin representations of messengers served as advertisements as well. In the window of Henry Miller's New York Western Union office was placed "a life-size cardboard figure of a bright, hand-

Figure 5.2. *Messenger mannequin as advertisement, 1917*

Source: *WU News* (March 1917), 167.

some-looking youngster, attired in the full regalia of the service." A 1924 photo of a PT window display shows that WU's competition had a stand-up full-sized messenger boy cardboard display as well, though theirs was not a photo but an illustration. And at least one WU office hung a full-body messenger dummy out the window as an eerie billboard[25] (see figure 5.2).

Print advertisements relied not only on messenger images, but on tales of messenger heroics. Press releases indicate the life-and-death importance that telegram companies wanted customers to associate with telegrams. The value of the telegram itself was tied to the hope that human messengers would do just about anything to deliver it. After all, how important could a message be if it was simply phoned? The telegram, delivered by messenger, on the other hand, was potentially momentous: "The messages he carries may involve staggering sums; lives may rest on his slender shoulders. Anxious people await his coming. Men of importance halt their deliberations at sight of the messenger, in natty uniform, to whom all doors swing wide."[26]

Figure 5.3. *Telegram as a messenger of tragedy, 1883*

Source: *O&E World* (April 7, 1883), 209.

This image of telegram importance was derived ultimately from the telegram's use as a death message, and from the messenger's duty to deliver such a message in person (see figure 5.3). Death messages were often described as requiring heroic messenger efforts. Two Houston, Texas messengers delivered a death message that arrived at 11:00 P.M. to a residence eleven miles away, for a special bonus of $2.50 each. They started out on bicycle, but rain and mud forced them to backtrack and go by train. It took them until 3:00 A.M. to find the recipient, "but by their labors a ten-word message had reached its destination in time to permit the man who received it to catch the early train for Illinois." This idea that a death message was important enough to warrant immediate delivery slid easily into the idea that messengers handled the most important and urgent of messages in general. As the WU slogan said, "Don't write—telegraph!"[27]

Heroic messengers were described as being able to find patrons in the most difficult of circumstances. One messenger delivered a telegram to a passenger who had already boarded a train. The messenger hopped the train, delivered the message, and got off at the first stop. Another messenger delivered a telegram to a moving barge by dropping it, tied to a rock, from a bridge. One messenger even had to "take off his uniform, strip to his union suit, and, with his message between his teeth, plunge into the water and swim out to the barge." And it was a messenger who delivered a governor's reprieve to a man sentenced to be hanged, on the night of his scheduled hanging. All of these stories built up the myth of the telegram just as they built the myth of the messenger.[28]

But messengers had more mundane obligations to telegram sales as well. They were often expected to actively sell the telegraph service, both on company time and on their own. Here was another reason that it was in the company's interests to pay a piece wage to its messengers: to motivate them to become little salesmen for telegrams and replies. In the 1930s, WU messengers had to carry around mimeographed slips of paper with them for reporting requests from customers: "Have moved," "Want call box," "Want box moved," "Need blanks," "Want charge account," "Want clock," "Are using Postal," "Want collect cards," "Have complaint," "Compliments our service," and "Want errand rates."[29]

Since telegrams were most often thought of as emergency communications tools, successful solicitors sometimes resorted to "ambulance-chasing" techniques. In 1894, a PT messenger from Birmingham, Alabama, solicited guests of a hotel where a fire struck, figuring they would need to wire for money. Another tactic was to solicit greeting telegrams to wives and mothers by embarrassing their husbands and sons in taverns. In 1938 congressional testimony, it was revealed that WU compelled messengers to solicit special-occasion greetings after working hours for holidays like New Year's and Mother's Day, setting quotas that had to be met. "The greeting quotas set by the company are usually so high that the messengers must solicit late into the night, often as late as 2 to 3 A.M., in order to meet these quotas. Beer gardens and liquor establishments are recommended by company officials as the best places to solicit. These messengers, who are in the majority minors, are often compelled by the customers to accept drinks in order to make a sale and thus many of them become intoxicated." Intoxicated or not, this worked out to twelve- to fourteen-hour days.[30]

The practice of ordering messengers to seek out telegraph customers had its roots in 1850, when Magnetic Telegraph president William Swain ordered that messengers with transmission blanks and pencils should enter train cars at every station to encourage passengers to send messages without leaving their seats. Former messengers from the 1940s said they still "hustled" outgoing messages on trains, walking through the cars with a stack of telegraph blanks and a rate sheet. In order to do this, a messenger had to know what the different categories of telegrams were, how much each cost over various distances, and how to properly count words. To simplify these tasks, a special 35¢ "travel message" was available, for fifteen words or fewer sent anywhere in the country. Such messages received top priority, even over regular full-rate messages, because above all else a travel message had to reach its destination before the sender did![31]

By 1930, WU was even giving out prizes to its top boy salesmen. Some fifty-two WU messengers were taken to Washington, D.C., on a trip as a reward for their sales during the preceding holiday season. Western Union "super-salesman messenger" Leon Richey, age sixteen, of Plainview, Texas, sold 279 holiday messages and won a free trip to the Texas state capitol at Austin. Two Atlanta, Georgia, messenger boys each sold 353 Thanksgiving messages and 333 Valentine's Day messages. And a messenger from Mineral Wells, Texas won a sales contest for selling 42 Father's Day messages.[32]

The more telegram sales hinged on the messenger, the more important the messenger's personal relationship with the customer became. In a very real way, when a customer sent a telegram, that customer was purchasing the personal care that a messenger boy would take with it. This is an important aspect of service work that differs from production work: the social construction of the worker's personality. Anthropologist Robin Leidner recently termed such work "interactive work," or work "where workers' looks, personalities and emotions, as well as their physical and intellectual capacities, are involved, sometimes forcing them to manipulate their identities more self-consciously than workers in other kinds of jobs." This was certainly part of the picture when WU customers were told in full-page advertisements to envision messengers and "think service."[33]

For example, even though the official rules forbade it, a good messenger's duty did not end with handing the customer the message. If the recipient was illiterate, the messenger might be asked to read the message out loud. And many messages were somber ones, especially for the young children who carried them. In 1878, one messenger boy wrote, "Hardly a day passes but that I have death messages to deliver, and several times I have been obliged to witness such a scene. . . . It is truly a messenger's hardest duty." A former messenger from 1934 recalled how he dreaded "the 'Two Star' death messages, which we delivered with the utmost caution and concern for the addressee."[34]

Death messages overwhelmed the telegraph system during times of war and disaster, requiring the telegraph companies to reform the image of the telegram on a regular basis. That, plus the gradual siphoning off of business communication to specific telegram spinoffs (including the telephone), caused the gradual rise in proportion of special-occasion greeting messages. The greeting-card industry in England had been around since at least the 1840s, but it wasn't until the 1870s that Louis Prang, U.S. developer of a multicolor printing process to duplicate oil paintings into what he called "chromos," began to sell colorful Christmas greeting cards in America. The practice waned in popularity in the 1890s, but in 1906 a number of U.S. publishers such as Alfred Bartlett and Fred Rust began to issue greeting cards again, this time with prewritten sentiments inside.[35]

Western Union saw an opportunity in this new social practice. In 1912, the company introduced a special Christmas holiday form for greeting-card telegrams, where "decorations served to embellish holiday sentiment, and helped to divorce telegrams from an unpleasant feeling they were used only to convey bad news." Not only special blanks, but special envelopes were used, to assure the consumer before opening the telegram that it was good news (and not a death notice). By 1939, a sender could select from some thirty-two different prewritten messages, sending the greeting telegram along with candy, books, or cigars.[36]

Now WU was itself a client of its own Dramatized Delivery service, in partnership with gift, candy, and flower retailers. By 1930, a customer could send a "telegram cake" baked by Gertner's Bakeries in New York City. Western Union had an official Fruit Telegraphic Delivery Service, where messengers would deliver baskets of fruit. Or a patron could send flowers using a florist who was a member of F.T.D.—the Florist Telegraph Delivery association, which used WU messengers. In 1936, new WU president Roy B. White even tried marketing "Kiddiegrams," telegrams decorated with nursery-rhyme characters for adults to send to children for 25¢ (20¢ for local-only delivery). This transformation of the telegraph system into a greeting service reached a peak when in 1925 WU urged consumers to send holiday gifts of wired money, not presents: "The Society for the Prevention of Useless Giving is reminding its members that this year the telegraph companies again offer to deliver for them brand new money to friends and relatives who have shown marked antipathy to white elephant gimcracks and geegaws." (This also meant that messengers worked on Christmas morning.)[37]

The shift from business telegrams to social greetings varied geographically. By the mid-1920s, business messages were declining in many areas of the country, and social messages may have been keeping offices in those areas afloat. In 1925 in Tullohoma, Tennessee, the majority of messages were social, as local businesses tended to use the telephone. And vacation areas were busy sites for social telegrams—not only a geographical variation, but a seasonal one. In 1929, vacation telegrams could be filed "from hundreds of other roadside stands, filling stations and garages throughout the country."[38]

Western Union analysts worked relentlessly to pin down the market for social telegrams at this time, coming to the conclusion in 1927 that "Jewish people send telegrams of congratulations and well-wishing much more frequently than members of any other group." They figured this out by tabulating messages sent on Jewish-specific occasions, such as religious holidays or marriages. Western Union carefully considered the needs of such markets: "All Jewish New Year messages will be delivered by Western Union unsealed, to conform to the rules of the Hebrew faith which prohibits the opening of a sealed envelope during the holidays."[39]

But no message epitomizes the shift to greetings like the famous "singing telegram." Western Union publicity manager George Oslin was credited by the company for inaugurating the service. Oslin wrote that in 1933, at the height of the Depression, "Western Union was losing a million dollars a month and was desperate for cash," especially because "people who lived during World War I still perceived telegrams as being synonymous with death messages," so he had a female operator deliver a birthday greeting in song over the telephone to popular singer Rudy Vallee. The next day, the *New York Mirror* reported the story, and apparently customers began requesting the service. That call was the beginning of an informal singing telegram service at WU: "Operators crooned them over the telephone and, later, uniformed messengers delivered them for an extra fee."[40]

Postal Telegraph quickly took up the idea as well, advertising, "Sing to her via proxy via Postal Telegraph messenger!" But WU officially disdained singing telegrams until 1938. One WU executive said, "Roaring love lyrics and popular tunes over the telephone to our valued clients, who, for all we know, are not in the mood for such attention, seems to us rather nonsensical." By 1940, however, the service was official, though only a limited amount of songs and lyrics were available. Messengers could deliver singing telegrams by this time, and these were often the best-paying duties for the boys: "I hit it rich when singing telegrams came along as [I] was in the High School glee club and got all the night business at 50¢ per message—sang at banquets and also wild nite club parties."[41]

It may be hard to believe, but singing a telegram was not the oddest thing a messenger might be called to do. Only a year after it opened in 1872, ADT New York City had already begun to respond to bizarre calls for messenger work. P. T. Barnum called for a boy to play euchre with him. One former messenger reminisced in 1904 about carrying a suitcase for a man so the man could sneak out of a hotel without paying the bill (though he paid the messenger). By the 1920s, WU could rattle off a long list of unusual services messengers had performed: fanning a "portly man" on a hot day; administering medicine to a sick man; collecting pennies from slot machines; guiding an eloping couple to a minister; supervising the moving of a household; and killing a mouse for a frantic housewife.[42]

Some of these duties verged on the absurd. Dog-walking messengers brought dogs into the office on rainy days and merely tied them to the wooden railing separating the operators from the customers for the duration of the "walk." In 1937, a WU messenger was hired by notoriously quirky inventor Nikola Tesla to feed pigeons in front of the New York Public Library twice a day at 10:00 A.M. and 3:30 P.M.—"he looks out for any that fall ill, and sees to it that they have their share of the five pounds of corn he distributes at a feeding." Anything for a tip![43]

There was a clear class component to such tasks. Odd jobs were advertised in order to promote the idea that to employ a messenger was to live a life of ease. In a footnote to her dissertation on the telegraph, Annteresa Lubrano suggests that "the use of these messengers mediated the effects of social status for some middle class families by affording them access to lifestyle services that the wealthier members of society have on a permanent basis." But whether or not the "middle class" generally used messengers in this way, the image of service for the affluent was important for the telegraph companies to maintain—Postal Telegraph even offered a "social secretary" service in the 1930s.[44]

All of these messenger services were meant to serve many purposes at once. First of all, the more varied the jobs messengers could do for customers, the more chances the messengers would have to sell telegrams. Second, the more heroic and trustworthy messengers appeared to be, the more important and safe the telegram would seem. And third, especially in the years following World War I, the more whimsy that could be added to the messenger service in order to shed the dismal image of the telegram as a death message, the better customers would feel upon receiving telegrams. Thus the push to sell greetings to consumers was not only an attempt to find a new market, or an attempt to find more things for messenger boys to do besides "loafing"—it was an attempt to redefine the telegram itself.

Both the transformation of business telegrams into advertisements and the transformation of social telegrams into greeting cards were linked to a conscious move on the part of the telegraph companies away from the image of the telegram as a death message—and the image of the messenger as a bearer of bad tidings. But a fundamental contradiction was at work in these campaigns that further undermined the effectiveness of the plain old yellow telegram.

The contradiction was that even though serious and somber messages kept consumers leery of telegrams, business use of telegrams for advertising played directly on these expectations of seriousness. "The only telegrams which many people receive are messages concerning death, marriage, birth, severe illness, and other unusual matters where speed and accuracy count more than anything else. So we have all come to feel that a telegram contains an important message, that it would be used only for an extremely important communication." A customer could hang up on an unsolicited phone call or throw away junk mail, but a telegram demanded immediate attention. If that were to change, the telegram would be no better than any other information commodity.[45]

But just when businesses were relying more and more on the image of a telegram as an important message, Western Union's desire to remake telegrams from somber notices into happy holiday greetings was undermining that very image of seriousness. In 1932, WU proudly asserted that "[t]he public has been educated to form a new conception of the telegram. Where once the telegram was regarded as primarily for emergency use, the public now understands that its social and business uses are numerous." They claimed, "Wires once tearfully received now bring smiles and pleasure." But now the telegram seemed frivolous. "The woman of the household does not faint at the approach of the messenger today, but wonders who is inviting her to a bridge or tea. In turn she sends her invitations by Western Union and they are delivered by messenger on the attractive social message blank."[46]

In the same way, Western Union's practice of advertising the telegram through unusual newsworthy events served to make the telegram seem unwieldy for day-to-day correspondence. In May 1927, on the occasion of Charles Lindbergh's famous solo Atlantic flight, WU offered a special "congratulations" telegram for sale. Customers sent 55,000 messages to Lindbergh; the telegrams were all loaded on a truck and paraded after him in Washington, D.C. "One telegram, from Minneapolis, was signed with 17,500 names and made up a scroll 520 feet long, under which ten messenger boys staggered." Who would use such a special form of communication for the mundane chores of everyday life?[47]

Western Union was aware of this contradiction, because when advertising telegraph use to business, the company alleged that the former fear of the telegram still remained in the mind of the consumer, but hidden as "a subconscious respect for the yellow blank":

> In the early days of the telegraph, telegrams were used only for extremely urgent purposes, such as accident, illness, death or some other calamity. In those days telegrams took on a character of impressiveness that has remained. In these days when a uniformed messenger appears at the door, very few people faint from shock, but there is still no man who feels safe in leaving a telegram unopened. There is still the thought that it may contain a matter of life and death, a business deal involving a large sum of money, or some other matter of great importance.[48]

Linking the telegram back to the uniformed messenger thus highlighted the seriousness of the communication, while defusing the customer's fear—or so the company hoped. Ironically, on the eve of World War II, WU would still be telling customers, "It has been a long time since telegrams were only used to convey BAD news."[49]

The telegraph companies had for many years promoted the idea that messengers were taken as seriously as the telegrams they carried. Western Union was fond of saying that only telegrams, hand-delivered by messengers to company officials themselves, made it through the mass of junk mail that was routinely processed by secretaries. One WU advertisement from the 1920s imagined "A red-headed office boy with a saucy pug nose" who sorted the mail: "Beside him was a waste paper basket with a maw that yawned like a hungry animal. Every few seconds, with a quick flip, he tossed something into the basket, remarking facetiously, ''Nother advertisement for you, old boy!'" This was the fear WU wanted to instill in businessmen: "It hurts your self-pride to think that an impudent boy of sixteen has the power to hold up your business message." But as the telegraph company proclaimed, "No office boy assumes to take liberties with telegrams."[50]

Western Union was countering fears of the "impudent" office boy with the image of its own responsible and diligent messenger boy. Another 1920s advertisement showed a silhouette of a messenger boy delivering a telegram to a businessman while other men waited in the lobby, saying: "Precedence. Past the crowd in the lobby—straight to the man they're waiting to see, go Western Union Telegrams, Day Letters and Night Letters. The yellow envelope gets the precedence everywhere. It delivers *your* message before the other fellow has the chance to shake hands." Of course, it wasn't "the yellow envelope" that went anywhere, it was the uniformed messenger boy, rarely discussed in these ads but nearly always present in the visuals. Not surprisingly, former messengers interviewed for this book confirmed that their business messages were usually just dropped off with receptionists, not delivered to powerful individuals. Even some of WU's own publicity photos belied this fact, showing messengers flirting with secretaries.[51]

Messenger boys appeared differently in advertisements for different kinds of telegrams—from the serious to the ridiculous, telegrams were always accompanied by a particular image of the boy. Western Union publicity shots showed polite messengers delivering telegrams directly to businessmen, jovial messengers singing holiday greetings or delivering flowers, and stern, speedy messengers running business or death messages to their locations lickety-split. In theory, messengers were never supposed to know what a telegram contained, but in practice they did. Death messages had two black stars on the envelope, signaling somber tact. Advertising messages were often sent out in bulk, destined for targeted neighborhoods, so messengers knew what those "telegrams" really said. And holiday wishes came in decorative envelopes, specifically to allay recipients' fears. So the messenger really was able to react appropriately to the news he was bringing—especially if he wanted a tip.

Serious business messages and whimsical consumer messages coexisted uneasily through the 1920s and the 1930s, with a Janus-faced messenger boy bearing each one. But with the return of war in the 1940s, the telegram could no longer be anything but a death message. Even before Pearl Harbor, the telegraph unions attacked WU and PT for the "nonsense services" they provided. A cartoon from 1940 showed a customer at the service counter of "Coast to Coast Telegraph Co." surrounded by signs advertising services: "Clocks for sale; bus tickets for sale; let us do your shopping; buy a greeting 25¢; money orders sold here; stock quotations; babies watched; dogs walked; football scores; surveys made; boys for hire; greetings sung; airplane reservations; packages delivered; travelers cheques." The employee behind the counter sighed, "Sometimes I wish we'd go back into the telegraph business." The cartoon wasn't far from the truth. Just three years earlier, *Business Week* had commended WU and PT for turning their offices into "information centers" and "aids to totally unrelated companies." Patrons could find Hartford Fire Insurance agents, or make payments to General Motors on their cars, or buy American Express travelers checks, or pick up and drop off packages with Railway Express.[52]

The problem with these services, according to the telegraph operator unions, was the involvement of messengers, which, operators argued, slowed the entire telegraph network down. Union operators, afraid that they would be blamed for telegraph slowdowns (and subsequently replaced with machinery), pointed the finger at messenger "nonsense services" (but not directly at messengers) in self-defense. At 1939 hearings before the Senate Committee on Interstate Commerce, a New York City WU branch office manager reported:

> We, for instance, collect for the *Saturday Evening Post* and the *Ladies Home Journal*. We deliver patent medicines. In addition to that we deliver airplane tickets and sell express orders, and we also are in the telegraph business. For instance, a messenger going out with both a message and something else, may be delayed with the one or the other. If a messenger goes out with a death message he may also have a number of patent medicine samples, and he is required to handle all these things in addition to the rest of the job.

After testimony like this, in 1941 the unions convinced the FCC to order WU to discontinue the "nonsense services," and to devote the entire messenger force to getting serious telegrams through.[53]

By 1945, even though it planned a drastic modernization of its switching plant and was committed to reducing the messenger force once and for all, WU was requesting reinstatement of these services. The division manager of industrial relations for WU argued, "When business is bad, we have to resort to things to which the union might object in order to increase revenue.... Singing telegrams are a source of revenue that can't be laughed off." The singing telegrams did come back by 1950—but by phone only.[54]

Through the early decades of the twentieth century, the construction of the messenger as a service worker (and as a commodity himself) changed dialectically with the construction of the telegram. The telegram was transformed more and more into a social greeting service as particular business functions (stock quotes, newspaper articles, and intrafirm communications) were taken over by specialized telegraph technologies. At the same time, the messenger was transformed more and more into a direct-marketing force as the telegraph company searched for more business uses for its telegrams. Efforts like day letters and night letters were attempts to control this demand; but efforts like holiday greetings and special promotions would create seasonal and daily peaks of their own. Through it all, the popular image of the messenger always affected the popular image of the telegram. What consumers thought about messengers, whether from stories in the telegraph company's advertisements or tales from the larger media, affected what consumers thought about telegrams.

Contrary to the standard story of the telegraph, messenger importance did not decrease linearly with mechanization. Messengers grew in both absolute numbers and as a percentage of all telegraph employees during the 1920s, a decade when automatic switchers and Simplex printers were being installed on all lines. The marketing of messengers for other duties was meant to attract customers to the telegraph system, for social telegrams, direct-advertising campaigns, and business telegrams. This marketing was enhanced by presenting the messengers as performing all sorts of useful jobs throughout the city. The next chapter will explore how such personal service often relied on the age, class, and gender of the messenger boy himself.

CHAPTER SIX

THE LIMITS OF GENDER, CLASS, AND AGE

I am afraid you can hardly be called a man. Still, you are not a woman or girl,
and I shall feel safer for having you here.

—Wealthy, older female patron to poor, young
messenger boy in a dime novel, 1899[1]

Why were the messengers constructed almost exclusively as *boys*? As detailed in
chapter 4, ADT's innovative technologies of labor were one answer. ADT was
the first telegraph company to put its messengers in crisp military-style uniforms and to put them through regular military drills. ADT also expanded the
messenger's duties beyond simply handling messages. Besides handling packages and door-to-door advertising flyers, messengers could also be procured on
an hourly basis. In this way, businesses might call for temporary messenger boys
when they ran out of regular "office boys," putting them to work at any number
of tasks. Some brokerage offices would call for eight to ten boys first thing in
the morning and keep them all day long.

All of these conditions and duties were linked to the ideals (and limits) of
proper behavior for young, working-class boys. Similar to the demographic of the
messenger force, the demographic of the business interests they served was overwhelmingly male (and white) during the century of commercial telegraph use from
about 1850 to 1950. Masculine ideals from the business world spilled over into the
training of messengers, who were instructed to display values of military honor and
precision, upholding secrecy and guarding private property above even their own
safety. These societal norms help to explain why from the late nineteenth century
to the early twentieth, most messengers were boys aged anywhere from ten to
eighteen, with the majority between the ages of thirteen and seventeen.[2]

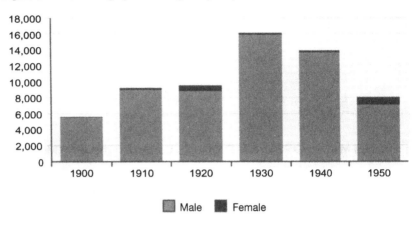

Figure 6.1. *U.S. telegraph messengers by sex, 1900–1950*

Source: U.S. Census (1900–50).

This is not to say that there were never any messenger girls. As early as 1886, while the New York City WU office was contracting for a force of 350 male ADT messengers, it also employed 33 female messengers of its own. Female messengers were almost always employed by the large national telegraph companies rather than the local district companies, resulting in a different geographic distribution. The official WU position as late as 1915 was that even in times of strikes, WU never employed girls as messengers. But this was untrue—messenger girls in small numbers around the United States were documented in numerous news reports, as well as in the census figures from 1900 to 1950[3] (see figure 6.1).

Generally, only in rural areas did messenger girls make up a significant proportion of the telegraph labor force. Mapping the states with significant percentages of female messengers from 1910 to 1950 reveals a pattern of female messengers in the midwestern and plains states, a stark contrast to the map of all messengers over the same time period (presented in chapter 2), which shows the bulk of messengers in the eastern states. This difference is another manifestation of the uneven telegraph geography between large cities handling rapid through traffic, employing large numbers of subcontracted messengers in a piece-wage relationship, and small towns handling slower way traffic, employing smaller numbers of "apprentice" messengers, often in a kinship relationship to a lone subcontracted operator[4] (see figure 6.2).

In urban areas, the employment of messenger girls was linked to extreme events, such as strikes and wars, which caused a shortage of young boys. From 1910 to 1950, the U.S. Census counted that women made up about 1 percent of the total U.S. telegraph messenger force, except in census years immediately

Figure 6.2. *Concentrations of female messengers by percent, 1910–1950*

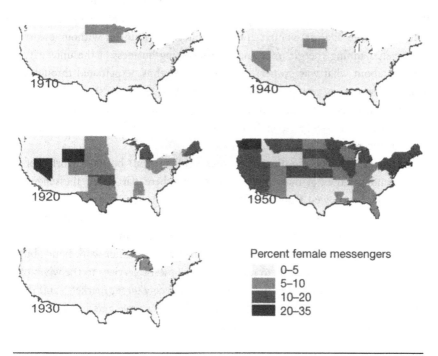

Percent female messengers
- 0–5
- 5–10
- 10–20
- 20–35

Source: U.S. Census (1910–50).

following world wars: in 1920 women held 5 percent of all telegraph messenger jobs, and in 1950 they held 10 percent of such jobs. In 1942, Western Union idealized these new wartime employees: "The average messenger girl is 17, a high school graduate and lives at home. She is 5-feet, 5-inches in height and weighs 110 pounds. She is athletically inclined, having been a member of her high school track, swimming or basketball team." A year later, the company claimed that "about 2,500 girls and women" served as messengers, indicating that adults served this replacement role as well, in so-called housewife agencies, where telegrams were dictated over telephone to suburban women in their homes for delivery to their neighbors at 10¢ to 20¢ each. By 1944, even Western Union's full-page telegraph advertisements featured messenger girls instead of boys. Interestingly, the company suggested that low wages for these female messengers were of no concern, because of the military crisis: "A number of women who have no economic need for employment are doing their bit by devoting their time daily to messenger work."[5]

Outside of these extreme situations of national crisis, any time girls were employed as messengers, controversy usually followed. For example, in 1903 WU messenger boys in Cambridge, Massachusetts, struck for a dollar per week wage increase. WU tried to replace the striking boys with girls, but Harvard University president Charles Eliot—quoted as saying, "I do not personally think it is very good policy to employ girls to do the work of men or boys"— enforced a rule against women entering college dormitories without escort, effectively banning the girl messengers from doing business at the university. An idea about what was proper "boy's work" was thus perpetuated through a legal limitation on women's urban mobility. This example shows how the assumptions of both the companies and their customers helped to maintain a nearly all-male messenger force from 1850 to 1950, even as the rest of telegraphy was increasingly feminized (and mechanized).[6]

Though the male-dominated (and masculine-ideal) business world certainly gendered the boy messengers who worked there, part of the messenger's gender identity flowed from a different source. American District Telegraph Company was originally conceptualized as a business service, but it quickly began to sell itself as a service for the home as well. Initially, this simply meant delivering commercial messages to businessmen at home after work hours; but eventually home service grew to mean selling domestic services to the wives of the businessmen as well. Presumptions about this new target market—and the "feminine" space it occupied—helped to gender the messengers just a bit differently than their adult male role models.

The telegraph companies advertised the social function of telegrams to women very early, and certainly messengers were expected to handle such communications when they arose. But these were by definition upper-class women with enough income and social obligations to require telegraph services. For such an affluent segment of feminine society, there were many different kinds of duties that working-class messengers could provide. One was security: even though a patron could often use her call-box to specifically request that a private "policeman" be sent to her residence, usually a boy was sent out instead, because of the frequency of call-box false alarms (and because of the higher cost of employing and dispatching adults). In 1873, the *New York Times* detailed several stories about messenger boys who helped thwart crimes, involving women who (allegedly) couldn't protect themselves. And as described in chapter 5, Dramatized Delivery was another service that could be marketed to homes as well as to businesses.[7]

By 1880, the editor of the *Operator* could rattle off a long list of domestic tasks that messengers performed for women: escorting them to and from "places of amusement"; walking children to and from school; searching for missing husbands at saloons, gambling houses, or brothels; accompanying drunk (and often broke) husbands home from such places; taking care of pets; delivering dinners, liquor, and groceries; paying bills; and pawning articles without a husband's knowledge. Such lists were repeated often over the next fifty years, usually tied to large cities like New York and Chicago.[8]

What such duties suggest is that a messenger's most important function for household consumers was to act as a mobile surrogate for immobile women. This blurred the messenger's gender position, combining unrestricted (masculine) access to the city (and especially to the poorer sections of the city) with domestic (feminine) activities of cultural patronage and home provisioning. As boys, messengers operated unnoticed, preserving the anonymity of their clients and avoiding embarrassment. Not only gender, but age and class are key in understanding this task of messengers, especially in the years before Progressive child-labor and mandatory-schooling campaigns began to raise the average messenger age above fourteen. On the one hand, young working-class messengers who performed errands and duties for older upper-class women were taking on a chivalrous man's role of provider and protector. However, in principle these same messengers were less of a threat to women (especially in the women's own physical spaces) both because they were only prepubescent boys and not yet full sexual beings, and because of their subordinate "servant" class position.[9]

The familiar notion of "separate spheres" for men and women in the turn-of-the-century city is useful here. Under this idea, men were expected to literally move about in the public world, participating in commerce, politics, and commercialized leisure, while women were idealized as remaining home to nurture children, to arrange family and social events, and to manage personal property and servants. Obviously such an idea had built into it certain aspirations to upper-class standards of living, as well as certain assumptions of native-white access to public amenities. The separate spheres concept is a convenient simplification of contemporary attitudes, and should be understood to describe an influential ideology of the time, not a set of actual sites, universal norms, or rigid boundaries. Still, for all its simplifications, the separate spheres concept is rooted in a reality of sex-based discrimination that lasted through the turn of the century, and thus it is a valid tool if used with caution. Instead of describing or explaining *fixed* spaces, the term should be understood as a way to explore the changing spatial *processes* by which actors conceded to and/or pushed the boundaries of those spaces.[10]

The separate spheres framework suggests that the messengers, as boys but not yet men, were able to move between the domestic and public spheres with ease, occupying a blurry gender position themselves. As children, they were literally "at home" in the woman's domestic sphere of the household, and were employed by those upper-class women who presumably would have held the separate spheres ideology most strongly; as prepubescent teens, they were considered appropriate escorts when those same women ventured outside in the public sphere, holding a man's place by the woman's side but not presenting a man's sexual interests; and as young men, their idealized future prospects in the

Figure 6.3. *Illustration from a messenger boy dime novel, 1903*

RING !

Source: George Ade, "Handsome Cyril, or the messenger boy with the warm feet," *The strenuous lad's library* 1 (Phoenix: Bandar Log Press, 1903).

world of business gave them license to move through the office corridors and back rooms of the man's private arena of the "public" sphere, the spaces of corporate capitalism and masculine entertainment.

This tense gender identity of the messengers is reflected in the dime-novel literature of the time as well—the stories that boy messengers themselves were likely reading. Most messenger stories relied on a series of linked tropes: the messenger is an abandoned orphan or the son of a single mother, living in lower-class poverty, yet he excels at his messenger job, and thus is clearly destined for greater things. His knowledge of the streets and his uniform allow him to move stealthily through the city: "his presence would not be likely to arouse any suspicion, where a man's would." Thus he falls into a plot where he is called upon repeatedly to serve and rescue helpless and immobile women: "Why, you don't know nothin' 'bout follering. Leave that to me. I'll shadow him" (see figure 6.3). And in the end, the messenger always reaps a surprise reward of both family and career, leaving his childhood working days behind for a respectable adult male life. Repeatedly in such stories, a not-quite-masculine messenger finally becomes a man—an ironic result, given that in the real world, critics of the "penny dreadfuls" argued that stories of "the wild behavior of dissipated boys in great cities" served to "poison boys' minds with views of life which are so base and false as to destroy all manliness and all chances of true success."[11]

This contradictory gender makeup for messengers helps to explain the wage and union position of the boys during this time period. From 1890 onward, office work in the information industries was becoming more routinized, more mechanized, and more feminized. However, "feminization" meant not simply the hiring of women, but the construction of jobs with low wages, few options for advancement, and high turnover rates, since the male manager's assumption about a woman's career was that she would work only in that short time period between the end of structured schooling and the onset of marriage and childrearing. From the point of view of capital, the messenger job, though idealized as mythically male, really bore more and more resemblance to the woman's dead-end clerical job. And even from the point of view of labor, it was easier for a female automatic operator to join the telegrapher's union than it was for a boy messenger to become a member (as described in chapter 9).

The employers of messengers thus had an interest in preserving this special position of the messenger. Relying on social ideals of masculinity, they could both sell messengers to male business customers and convince young boys to apply for messenger work. But with turn-of-the-century social ideals of femininity came a chance for messengers to sell services to female domestic customers as well. This paradoxical gender position gave messengers open access to nearly all the spaces of the city; however, it also carried the seeds of the only legal limitations ever put on such messenger access.

The battle for limits to messenger mobility in the city was rooted neither in the man's sphere of the capitalist business world, nor in the woman's sphere of the well-kept home, but in an urban sphere arguably both male and female, both domestic and commercial: the liminal space of "vice." In turn-of-the-century New York City, writes Luc Sante, prostitution was ubiquitous, and this limited women's mobility: "There were men on the street, on public conveyances, at places of amusement, who could spare a dollar or two for a rapid sexual fix. Any woman by herself was fair game, and two together might be thought a team. Any woman out after dark would be assumed to be a whore." But Sante notes that the mobility of prostitutes was also restricted, describing "a young woman who entered a brothel, where her earnings would be taken by the madam, who would pay her only a meager allowance, and where her movements and activities would be as closely monitored as if she were in a nunnery." Into this space came the messenger, and the responses of both reformers and employers illustrate that where the "separate spheres" collided, ideas about the proper limits of gender, age, class, and work were put to the test.[12]

Before the the turn of the century, New York prostitution revolved around brothels, usually run by "madams." The clients were "sporting men," according to historian Timothy Gilfoyle: a subculture, apparently cutting across classes but most popular among educated, middle-class males, that rewarded "autonomy, promiscuity, extramarital sex, and physical isolation from the nation's strict Victorian mores." But after the 1896 Raines Law banned saloons from serving liquor on Sundays, new "Raines Law hotels" (really just saloons with a few beds in back) became more profitable locations of prostitution than brothels. With the prostitution hidden inside, and the hotels built close to theater districts, there was less spatial segregation of the vice industry: "In contrast to the prescriptive Victorian literature of the era, social elites shared the streets and institutions of the neighborhood with more ribald elements of New York's sexual underworld."[13]

In 1905, a prominent new reform group, the Committee of Fourteen for the Supression of the Raines Law Hotels in New York City, began an anti-prostitution campaign in New York, and won the Ambler Law, which closed down most Raines Law hotels. Many other groups supported the cause: municipal reformers wanted to clean up the vice districts and remove corrupt officials who condoned them; urban doctors wanted to stop venereal disease; Christian moralists wanted to stamp out sin; and contemporary feminists wanted to free women from sexual bondage. In particular, between about 1909 and 1914, tracts warning of the "white slave trade" in prostitution were everywhere.[14]

Historian Barbara Hobson argues that by the Progressive era, prostitution had become a complex, rationalized operation, since it was "one of the few businesses that requires little capital, offers quick returns, and involves few risks for those who are not doing the actual selling." Thus, "a greater division of labor was evolving in sex commerce: there were proprietors, pimps, madams, runners, collectors (who paid off police), doctors, clothing dealers, and professional bondsmen." Historian David Nasaw adds that young "newsies" who delivered papers to prostitutes might also have arrangements to deliver food as well.[15]

Messengers had long been part of this division of labor; as early as 1883, the *Electrical World* noted that Pittsburg messengers were "exposed to vicious influences" from the "houses of ill-repute" where they ran "questionable errands" (though the editors concluded that it was "the duty of the parents to take care of their sons' morals"). By the turn of the century, red-light districts were so common in cities that printed "sporting guides" and "blue books" listed brothels with descriptions, locations, and prices; in this context, messengers were simply the most up-to-date of a variety of printed and oral sources of information on urban vice. Kickbacks, large tips, or outright overcharging for service provided reason enough for boys to work these areas. And steady profits provided motivation for telegraph managers to dispatch the boys there; in 1888, the ADT Philadelphia annual report to the president lamented, "It will be noticed that there is a falling off in some of the districts, which may be caused by the great reduction of saloons under the Brooks License Bill and houses of ill repute having to close and seek new quarters."[16]

Although street boys had been stereotyped as agents of vice as early as 1872 with Charles Loring Brace's exposé *The Dangerous Classes of New York*, public attention was focused on this aspect of messenger work through the efforts of "muckraking" journalists, starting with Jacob Riis in 1892. In his book *The Children of the Poor*, Riis argued that messenger service "ought to be prohibited with the utmost rigor of the law" because of "the kind of houses they have to go to, the kind of people they meet, [and] the sort of influences that beset them hourly at an age when they are most easily impressed for good or bad." Similarly, John Spargo's 1906 book *The Bitter Cry of the Children* said of the ADT boy who worked in red-light districts, "He smokes, drinks, gambles, and, very often, patronizes the lowest class of cheap brothels. In answering calls from houses of ill-repute messengers cannot avoid being witnesses of scenes and licentiousness more or less frequently. By presents of money, fruit, candy, cigarettes, and even liquor, the women make friends of the boys, who quickly learn all the foul slang of the brothels."[17]

Such situations were interpreted in one of two ways by contemporary reformers. Some argued that the messenger service merely attracted boys who were already prone to vice: "Many boys who are unwilling to work steadily are attracted by the irregular hours of the messenger service, and by the opportunities it offers to loiter on the street and to indulge in petty street crimes." Others thought that the messenger service itself caused vice in impressionable boys: "They become part of the mechanism of vice. And then they become part of vice itself. It is a process of inevitable absorption. The disease of vice penetrates the whole body of the street service among boys."[18]

But common to both perspectives was often a new kind of gendering of the telegraph messenger's job: by day, the masculine boy worked for respectable businessmen; by night, the feminized messenger worked for disreputable prostitutes. And though clearly the telegraph company was driving this grab for "the profits of vice," it was the women in the messenger's life who were thought to bear responsibility for the situation. Night messenger service for vice turned the vigorous boy messenger into a "slouching" and "shambling" creature. Set in a context where physical vitality was a step toward proper masculinity, this meant that night messenger service was a feminizing influence. And if feminized boy messengers were the victims, then adult women were to blame. Not only the female prostitutes who were the customers of the telegraph system, but all women who employed messengers were apparently at fault: "From carrying messages for the women of the town they go on to carrying cocaine and other drugs for them." Or perhaps the mothers of the messengers themselves were at fault, because as one author declared, "there has never been much evidence of maternal care in the life of the boys who spend their working hours carrying messages and parcels almost exclusively among disreputable characters" (except for a case he described where a madam had her own son work as a messenger).[19]

Such sensational language was a call to arms for the newly formed National Child Labor Committee (NCLC). The NCLC's Ohio Valley secretary, Edward Clopper, who was already investigating "street work" in Cincinnati for his own Ph.D. dissertation, focused his attention away from newsboys and onto night messengers. The NCLC's director, Owen Lovejoy, ordered a two-year national investigation of night messenger boys in twenty-seven cities in nine states, involving thirty messenger companies, including Western Union and ADT. Urban photographer Lewis Hine was hired by the NCLC to turn his lens to messenger boys as well. The familiar Progressive pattern of journalistic notice spawning social survey leading to regulatory reform had been set in motion. The NCLC's conclusion, buried in their "unprintable" report of 1910, was that "The messenger's cap is an open sesame to the underworld."[20]

But even though the underworld was a varied terrain, with countless male spaces of drinking and gambling, the reformers fixed upon the spaces inhabited by female prostitutes. Clopper reported, "The boys get them chop-suey, chili-con-carne, liquor, tobacco, opium, medicines and articles used in their trade, deposit their money in the bank and one instance was found in which a boy was actually required by a prostitute to clean up her room and make her bed." Lovejoy added that the boys were "used as agents for the purchase of cocaine, opium and other narcotics, for purchasing liquor during lawful and unlawful hours, for the purchase of drugs used to render insensible the patrons of the disreputable house that they may be robbed by its inmates, to carry notes from the cells of arrested women to their male companions informing them of their plight."[21]

In reality, of course, even in sites of prostitution, messenger work was serving men as much as it was serving women—not just "johns," but pimps and protectors, bar and hotel owners, police and politicians on the take, and other go-betweens. Yet reformers feared that the specific association with female prostitutes would corrupt the messengers, in four main ways. First, in a sort of guilt by association, messengers—especially the youngest among them—might actually become friends with those engaged in vice. Lovejoy claimed that messengers were "frequently hired by the hour to serve the food or drinks they have been sent to purchase, in sitting-rooms occupied by the most vicious characters." The mere contact with criminal behavior was thought to have a disease like effect on the character of the child. But Nasaw points out that in many neighborhoods "Prostitutes were a part of the life of the streets. The children played with them, teased them, and ran errands for them. There were no secrets—and very little shame—between the groups. They shared the same public space and had little choice but to get to know one another."[22]

The reformers' second fear was directed at slightly older messenger boys: once messengers were exposed to the inhabitants of the underworld, they might join them as criminals. Clopper often argued that "the largest number of delinquent boys is found in those occupations in which the nature of employment does not permit of supervision, namely, newspaper selling, errand running, delivery service and messenger service." He feared that such messengers would "become saloon keepers because they have become well acquainted with this method of making a livelihood," or would be "attracted by the life of 'ease' which opens before them and enter into agreement with prostitutes." Clopper quoted one messenger as saying, "You don't learn anything in the messenger business except to knock down [overcharge a patron] and to go around with prostitutes and gamblers. It kills a fellow." But this boy also said he made $40 to $75 per month including his generous tips ($10 to $18 per week).[23]

Ironically, from the reports of messenger mischief that survive in the archival record, it was not contact with the nighttime "underworld" that led messengers astray, but contact with the upper-class society of capital and privilege that messengers served openly by day. As telegraph companies increasingly depended on their lucrative money-order service, journals like the *Telegraph Age* lamented the "undesirability of trusting money to the messenger boys." The scrapbook of Western Union "messenger detective" E. C. Brower, compiled from 1897 to 1902, detailed messenger thefts of cash for money orders, women's jewelry, and bank deposits. One sixteen-year-old ADT messenger even stole $290, allegedly to fit into the high-society crowd that employed him to carry messages to exclusive clubs. Such crimes arguably had more to do with the low pay, harsh conditions, and limited hope for advancement in the messenger job itself than with any link to the "Red Light."[24]

The third fear of the reformers was that messengers would become sex customers themselves, a fear based on assumed gender differences in the sex drive. For example, settlement-house pioneer Jane Addams, following the theories of G. Stanley Hall as outlined in his 1904 book *Adolescence*, believed that a boy's "sex impulse" had to be channeled productively, lest it take over: "Every city contains hundreds of degenerates who have been over-mastered and borne down by [the sex impulse]; they fill the casual lodging houses and the infirmaries. In many instances it has pushed men of ability and promise to the bottom of the social scale." Addams argued that through clubs and playgrounds, such impulses could be productively defused, but "To fail to provide for the recreation of youth, is not only to deprive all of them of their natural form of expression, but is certain to subject some of them to the overwhelming temptation of illicit and soul-destroying pleasures."[25]

In this way of thinking, female prostitutes were seen not as victims in this underground economy, but as temptresses and corruptors. Revealing that "a large percentage" of the messengers whom he spoke with had venereal diseases, Lovejoy concluded that "the deliberate design of women thus to increase their patronage" was at work. Prostitutes were seen as entrapping "the most promising boys" because "these women give tips with a liberal hand," as well as gifts of liquor and cigarettes. A Chicago report on "The Social Evil" said of one fifteen-year-old messenger, "whenever he comes into a house of prostitution the girls fondle him and nearly always kiss him. At different times he has had sores on his lips."[26]

This fear of prostitutes mirrored contemporary fears of other urban women as well. Domestic servants (usually working-class women) were often feared to be teaching the children of their employers sexual and immoral behaviors, "poisoning the minds of the young and innocent and initiating them into habits of vice," according to one 1892 physician. Even female elementary school teachers (90 percent of all U.S. elementary school teachers in 1910) were feared to be emasculating boys through their attentions, argues historian Michael Kimmel: "Many men believed that cultural feminization was the direct result of the feminization of American boyhood, the predominance of women in the lives of young boys—mothers left alone at home with their young sons and teachers in both elementary and Sunday schools."[27]

In part a reaction to such fears, in 1910 the Boy Scouts of America was formed as a strenuous group run by men and for (future) men. Founder Ernest Seton thought that women's influence on boys at home was damaging, potentially leading to "the boy who has been coddled all his life and kept so carefully wrapped up on the 'pink cotton wool' of an over-indulgent home, till he is more effeminate than his sister, and his flabby muscles are less flabby than his character." The character reference is important here, for the argument was that the female influence could have a direct effect on a boy's susceptibility to vice. A 1914 *Education Review* article linked madonna to whore when it warned of the "feminized manhood, emotional, illogical, non-combative against public evils," that was resulting from the preponderance of women teachers in the grade schools. The irony is that such fears of women's influence were projected onto messengers who actually worked all day long under the ostensible supervision of telegraph *men*.[28]

From being weakened to the point of consuming the sexual favors of the prostitutes, it was a short step to the reformers' fourth and final fear: the general consumption of urban vice by the messenger boy. Nasaw notes that for working children of this time, "money bought pleasure *and* a place in the city." But messengers not only had a disposable income to spend in that city, they had a special knowledge of the city and mobility within it. Again, here was a fear rooted in access to space—but spaces of consumption, not of service. Addams worried, "Never before have such numbers of young boys earned money independently of the family life, and felt themselves free to spend it as they choose in the midst of vice deliberately disguised as pleasure." Reformer Ernest Poole, while at the University Settlement in New York, specifically pointed to the irregular piece wage of the messenger in this regard, and to the messenger's dependence on tips, saying that both led to "reckless spending" in gambling: "In one of the large messenger offices on Broadway it is common for boys to lose the entire week's earnings in the hall and stairway before reaching the street."[29]

This generalized fear about consumption was focused on a particular space: the saloon. Reformers' worries about saloons mixed together all three of their other fears about messenger boys—association with vice, adoption of criminal habits, and ruination through prostitution. This was not an irrational concern; prostitution at this time was somewhat dependent on the saloon, especially as brothels became more scarce. According to historian Jon Kingdale, New York City had over ten thousand licensed saloons by 1915 (one for every 515 persons), many in working-class districts. Prostitutes would be allowed in as long as they could attract men to the place and keep them drinking.[30]

But saloons were not places where messenger boys would be strangers. After all, boys aged fifteen and above were considered of "adult" saloon age, according to Kingdale, who argues that the working-class neighborhood saloon of 1890 to 1920 "was a neighborhood center, an all-male establishment and a transmitter of working-class and immigrant cultures." More important, saloons served cheap lunches, cashed checks, lent money, served as a mailing and message-receiving address for many a worker, and provided the only public toilets around. All of these functions would have been useful to the messenger on the job.

But reformers ignored the fact that saloons, hotels, and brothels were sites of labor for messengers, and thus sites of sales for the telegraph companies. In considering the messenger boys as children and not as workers, the solutions that the reformers proposed rarely indicted the telegraph companies directly. Instead, they attacked only the specific duties that brought messengers and vice workers together.

This was a somewhat surprising response. Hobson argued that many women's activists saw a broader meaning in the rationalized prostitution trade: "prostitution became the symbol of a corrupt male polity that sanctioned sexual exploitation and permitted vice syndicates, politicians, and police to profit from the sale of women's bodies." The telegraph companies profited too—part of this corrupt male polity. But there were few suggestions made that perhaps WU should remove itself from the vice business altogether, except for the wry observation of the *New York Times*: "There is some difficulty in seeing why the work complained of should be done at all, or why it should be much less injurious to men who had just attained their majority than to boys who were approaching it." In other words, if messengers were buying "knock-out drops" and running errands for criminals, that should be stopped outright, not simply made the work of adults.[31]

Considering the nature of the telegraph system, it is also surprising that technological solutions to this problem were not pursued. Some advocated prohibiting WU or ADT from "installing or permitting to remain in such places telephone, telegraph, or other means of communication." There was precedent for such a move, since in 1904 WU had agreed to "discontinue forthwith the collection and distribution by this company of horse race reports" because such reports were telegraphed to illegal pool-rooms. This time the company argued that if its call-boxes were removed from the red-light district, customers could simply phone for the messengers. But with the 1909 takeover of WU by AT&T (described in chapter 7), certainly both services could have been restricted—after all, the phone company itself advertised "One System, Universal Service." But neither the phone company nor the telegraph company was in the business of removing lines from profitable customers, so a technological fix for the vice problem was not to be.[32]

The NCLC's legal goals fell into a familiar pattern. New York state legislators had a history of spatially segregating children from vice, from an 1859 law banning unaccompanied children in performance halls to an 1892 law to keep children from performing in theaters, both passed out of fear of prostitution. The goal of the NCLC was a temporal segregation—their bill prohibiting the employment of messengers under-21 between 10:00 P.M. and 5:00 A.M. was easily passed. According to reformer Florence Kelley, the "Murray Night Messenger Bill" passed solely on the basis of the 1910 NCLC report, "without any opposition at all." No boy under sixteen could deliver messages, telegrams, or merchandise after 7:00 P.M.; and no boy under twenty-one could do so between 10:00 P.M. and 5:00 A.M.. The penalty to the employer was a fine of $20 for the first offense; $50 for the second; $100 for the third; and then imprisonment.[33]

The messenger laws were weakly enforced, but they did have an effect on the telegraph companies, who now had to risk the consequences of bad publicity. The NCLC kept up public pressure all the way up to World War I, even producing a poster entitled "Moral Dangers," which compared a photo of a child operating an unshielded circular saw to a photo of a bicycle messenger "carrying notes from prostitute in jail to the 'Red Light'" and claimed that "street trades also are hazardous because they expose children to moral dangers more deadly than circular saws."[34]

Western Union dealt with the constraints on its youngest messengers by tapping into another source of low-wage labor: the elderly. At first the company resisted this move, since they employed children for many specific reasons, such as low wages, abundant labor supply, and simple discipline tactics. But Lovejoy himself had suggested that perhaps "cripples, elderly persons, industrial misfits, and others beyond the probability of being tempted to wrong-doing" should be employed instead of young boys, noting that these workers, too, would accept low wages.[35]

Florence Kelley thought the new messenger law had a dramatic effect:

> I have never seen such an astounding change in the personnel of any occupation as the change from the slow, shuffling, shabby, irresponsible, little twelve-year-old boys sent to my office a dozen years ago when I rang for a messenger, and the somewhat shabby, stooping, but alert and eager white-haired men who come now to my office either by day or by night when I ring for a messenger. The pay is so poor that men of adult years do not willingly take it. Only the aged, and the irresponsible youth are to be had for the money that these corporations pay.

Kelley was not alone in her assessment. By 1918, the *New York Times* could report that night messengers were "mostly old men."[36]

Census numbers do support the contention that WU hired more "aged" messengers. Defining for a moment "elderly" as aged forty-five and over (since WU policy was not to hire any employees over age forty-five), it appears that in 1910, only 1 percent of New York City messengers were "elderly," but by 1920, 9 percent were. Census breakdowns don't reveal exactly what proportion of the messengers met the reformers' criteria for adulthood (being over age twenty-one), but they do show the opposite end of the spectrum: the number of messengers in the youngest cohort. Defining "child telegraph messengers" as those aged ten to fifteen, from 1910 to 1930 the proportion of "child" messengers did shrink: from 50 percent of all messengers in 1910 to only 21 percent of all messengers in 1930 (total numbers shrank as well). However, with children defined as aged ten to seventeen, no such decrease appears. Thus even with more of the oldest messengers and fewer of the youngest, the practice of employing teenagers continued[37] (see figure 6.4).

As time went on, and as it became clear that the NCLC law would be weakly enforced, WU shifted to a strategy of denial. Only five years after the New York state law against night messenger work was passed, WU was called to task by the Industrial Relations Commission for its continuing involvement with vice. Western Union president Newcomb Carlton testified that when mes-

Figure 6.4. *U.S. telegraph messengers by age, 1900–1950*

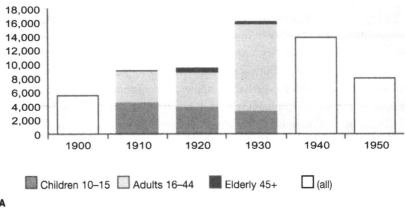

A

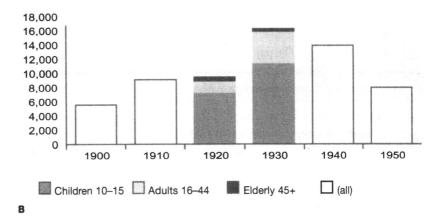

B

A: "Adult" defined as age 16 to 44. B: "Adult" defined as age 18 to 44.
Source: U.S. Census (1900–50).
Note: Blank years indicate missing data.

senger boys were sent to "places of questionable reputation," they were to be over twenty-one, and he denied that WU maintained call-boxes in the red-light district at all. But New York City messenger manager Belvedere Brooks said that WU sent boys wherever they were needed, and didn't keep tabs on the addresses: "the authorities should do that, the police authorities." Two messengers also gave testimony at the hearings. George Daly, fifteen, had made $7.50 per week working a 2:00 P.M. to 8:00 P.M. shift for WU in 1914; his friend Eli Rosenberg, seventeen, also worked for WU, on a 2:00 P.M. to midnight shift, for $30 per month. Both talked of responding to telephoned calls coming into the telegraph office asking them to purchase opium for customers for 25¢. When the commission announced its findings, it said of such messengers, "Their own morals and health are not only contaminated, but they disseminate the information thus gained to the boys with whom they are associated, and thus form an ever widening source for the contamination of the youth of the cities."[38]

Metaphors of "contamination" aside, it is clear that messenger boys quickly learned from each other on the job that the vice district was a source of steady and lucrative business. Reformer Scott Nearing told of a Philadelphia night-shift messenger who had been hired to wait in line all night to buy theater tickets that were to go on sale the next morning—a clear case of "proper" work for an upper-class client. Around 1:00 A.M., the boy was asked by two marines to show them to a "place" that hadn't been raided recently. The marines paid him 50¢ and since the woman of the "place" gave him 50¢ for each customer he brought around, he made a total of $1.50 on the endeavor. The boy reported that in night work, "you get sent to worse places, and you've got to carry heavier loads." He added, "there's lots of tips on the night shift, tho, and lots of boys likes to get the job." He said that even though he was under sixteen, he looked older, so age wasn't a problem. In any case, he would simply need an affidavit from a parent swearing he was sixteen, which could be had on the street for 25¢ to 50¢.[39]

Western Union's response to such stories through the 1910s was twofold. First, they countered the fears and tales of messengers gone wrong with stories of exceedingly moral messengers. The telegraph companies had long claimed that their messengers were honest beyond suspicion. But now the advertising stressed not just honesty, but heroics. Messengers all over the United States were saving children from burning houses, foiling robbers by leaping off their bicycles and tackling them, or stopping runaway horses. Here was a level of heroism above the "catching a customer on the train" antics described in the previous chapter.[40]

Second, WU began to make links with organizations such as the YMCA and the Boy Scouts, so they could claim that they offered wholesome leisure activities to their boys as well as instructive employment. The telegraph companies did everything they could to receive a stamp of approval from local boy's groups who could once again vouch for the "manliness" of the youthful messengers. By 1931, these two strategies were so common that WU could brag about their fifteen-year-old messenger from Jacksonville, Florida, who had thwarted a theft from a WU safe, and was also a member of a Boy Scout troop made up of WU messengers.[41]

Though such advertising certainly helped to polish the messenger image in the public eye, the issue of messenger vice soon faded from public view anyway, for several reasons. One cause was the increased employment of messenger girls with the coming of World War I. The war offered WU its first chance to publicly promote its female messengers without worry of moral outrage. From 1917 to 1918, WU hired messenger girls to replace young men going off to war, even in large cities. The girls wore dresses instead of the boy's uniform, but usually wore the messenger hats and sometimes even rode girls' bikes. Western Union put the most positive spin it could on messenger work for these girls: "they find their present work very healthful, as they get much outdoor exercise, which, of course, is conducive to good health, and good health means happiness, and unquestionably good looks as this they all possess."[42]

But the best spin WU could put on the situation was to point out that such girls were not "compromised" by their association with the messenger service, a tactic that implicitly spoke to the reputations of the messenger boys as well. After all, in 1903 the operators' union had charged messenger girls with prostitution themselves, "the young and innocent ones associating with older and less innocent ones, who in many cases have used the messenger service as a cloak for ulterior purposes." By 1920, WU could argue that "we have had representatives from the juvenile court and other numerous child labor organizations of this city to inspect our plant, and to thoroughly look into the safeguards we have thrown around our employees, and I think in each instance we have been given a clean bill of health." A sympathetic witness for WU agreed: "there has been only one complaint of any girl being subjected to any improper approaches while in the employ of his company, and that person has been committed to St. Elizabeth Asylum, he being insane."[43]

Of course, WU tactics weren't the only reason that the vice issue faded. As will be shown in chapter 8, the focus of child-labor reform actually moved away from vice issues to issues of the school-leaving age, where reformers knew that by keeping kids in the classroom they were automatically keeping them off the streets. Another cause was the changing nature of transportation in messenger work. As automobiles began to replace bicycles, WU had to hire adult men with licenses anyway, so it could just as easily use them to service the controversial areas of the city, especially late at night. And finally, with the help of aggressive military campaigns against sexually transmitted diseases, "red-light districts in hundreds of cities were shut down by 1920." Clearly the trade merely dispersed, as did the liquor trade after Prohibition during the same period. Nevertheless, prostitution did become less of a visible "blight" on the city, and thus less of a public relations problem. So did the messenger boy vice issue.[44]

Thus by the mid-1920s, WU was once again proudly advertising boys, not men, as its messengers. Future author Henry Miller, still a messenger manager at this time, later wrote that in the window of his office was "a life-size cardboard figure of a bright, handsome-looking youngster, attired in the full regalia of the service." According to Miller, "this piece of bait served two purposes: it

Figure 6.5. *Lampooning elderly messengers and celebrating vice, 1924*

Copyright, 1924 (New York World), Press Publishing Co.
"I'm a tellin' yer, Yer'll never succeed in this business unless yer 'tend to yer job."

Source: *PT* (June 1924), 39.

pretended to persuade the idler and the nitwit that in the service of the telegraph company there was ever open a glorious career; it also helped to break
down an erroneous popular conception. All messengers, it seemed to say, are not
idiots or septuagenarians." A *New York World* cartoon from 1924 made the same
point, lampooning anyone who would try to make a career out of the messenger service, showing an elderly messenger chiding two young craps-shooting
messenger boys by saying, "Yer'll never succeed in this business unless yer 'tend
to yer job"[45] (see figure 6.5).

Similarly, with the "underworld" a less visible component of the urban
landscape, the messenger's knowledge of the city was no longer being touted as
a sinister pass to the underworld, but as a crucial community service. A 1924
publicity photo of a uniformed messenger boy carrying a suitcase and accompanying a lady down the sidewalk was captioned, "The telegraph messenger
knows all the streets of his city, and knows everything else about it that is worth
knowing. Therefore he is a splendid guide for visitors."[46]

Though the attention to vice faded, the practice of serving vice apparently
continued. One former WU messenger who worked in 1933 in Deadwood,
South Dakota, as a teen, six days a week for $25 per month, remembered, "Most

of the tips which I received were from the red light district." Another former messenger from Long Island, New York, remembered how once in 1946 "a hooker offered herself as a tip," the same year that the official WU messenger manual proclaimed, "No assignment is accepted for him which is objectionable, undignified or which would tax his ability to fulfill."[47]

Dignified or not, there were new reasons for messengers to enter the spaces of vice as the telegraph turned from a necessary instrument of business into a domestic greeting-card service. As described in the last chapter, WU compelled messengers to solicit special-occasion greetings after working hours for holidays like New Year's and Mother's Day, setting quotas that had to be met. One messenger confirmed that this was still the practice in the 1930s: "Saturday night prior to Mothers' Day we were given blank copies of Mothers' Day grams along with various greetings that could be inserted if the customer couldn't come up with one of his own. The order was given, 'You guys are to hit every bar in Richmond Hill and sell them drinkers a Mother Day Gram to be delivered tomorrow.' The deal worked and many greetings were sold to men with tears in their eyes. I still don't feel right about having done that." Perhaps the most important conclusion to draw here is that the messenger was present in the spaces of vice precisely because those were also spaces of consumption, spaces of business, and spaces of capital accumulation in the modern American city.[48]

Masculine gender definitions of duties and working-class hopes of future careers, based on the idea that a working boy would soon grow into a working man, were key to creating a messenger force that could function effectively in the public business sphere of the city. But more feminized gender definitions, linked not only to class but to age and motherhood, were key to extending messenger work into the woman's domestic sphere at the same time. In this way, messenger boys occupied a blurry gender position themselves, which enabled them to perform both masculine and feminine tasks in both the masculine and the feminine spaces of the city.

Age and class were key in maintaining this gender position, because the mythical future opportunities of the messenger were intended to compensate for his present subordinate status. In a study of present-day female fast-food workers and male insurance salespersons, Robin Leidner shows how job duties that might be linked to one sex could be embraced by members of the other sex, by focusing on the gender-appropriateness of other aspects of the job. As will be shown in chapter 8, both the messenger boys and their customers could rationalize duties that might otherwise be considered demeaning for a young man—minding children or fetching women's toiletries—by accepting the assurance of the telegraph companies that the messengers of today would surely grow into the businessmen of tomorrow. Messengers who didn't buy into this myth merely quit the service—and they did so in droves. But messengers who stayed on took an active role in finding the urban spaces where they could reap the greatest economic gain, regardless of the actual "business training" that such places offered. This search for tips led the mobile messengers first to brokerage houses, then to sitting-rooms, and finally to bars and brothels.[49]

If the standard story of Progressive anti-vice campaigns is one of legislating and punishing women and not men for sexual promiscuity, then the messenger boy story is familiar yet somewhat different. The child-labor reformers of the early twentieth century believed that the messenger boy's unique gender position, based on age, class, and urban access, made him a vice risk. But rather than directing efforts at fundamental change in the public business and political sphere that enabled vice in the first place, reformers working on the night messenger problem tried to delineate and segregate a third sphere of urban vice, a world where men purchased entertainment that women produced, and where the female worker was to blame for the moral fall of the male consumer. In such a world, the male messenger serving female needs was simply a young child—only an adult male would be tolerated in such a job. What today might be termed a "communications decency act" for telegraphy resulted in the first actual spatial and temporal urban limits put on messenger boys, as well as the first substantial employment shift from the messenger boy to the messenger man. Until attention to the issue faded, the messenger boys could push these boundaries of age, class, and gender no further.

CHAPTER SEVEN

BOUNDARY WORKERS IN AN INFORMATION INTERNETWORK

We believe that the future development of the wire system in the United States will afford facilities for the annihilation of both time and distance by the general use of electrical transmission for written or personal communication, and will afford electrical communication of every kind of intelligence from everyone at every place to everyone at every other place.

—AT&T *Annual Report*, 1910.[1]

In actuality, each individual is the center of a converging web of communication media, which in turn extend his touch to all the world.

—*Communication Agencies and Social Life*, 1933.[2]

In 1901, Edward Calahan, inventor of the call-box some thirty years before, felt that the district messenger system was in no danger of dying out due to competition from the telephone: "We shall rather see it developed and its utility even more strikingly brought out than in the past." Sure enough, messenger employment grew right up to the Great Depression, through a period of intensifying competition between telegraph, post office, and telephone. The persistence of messenger labor in the telegraph industry is explained in part by considering telegraphy's complex relationship to those other information networks. Not only competition, but also cooperation characterized the tense relationship between the telegraph, post office, and telephone during the early decades of the twentieth century. In many ways, the three networks operated together as a single "internetwork." On the surface, consumers would choose which system to use for a particular purpose; but underneath, all three systems relied on each other for actual delivery.[3]

This internetwork manifested a particular sort of shared geography. For example, the telegraph and telephone networks shared wires, one leasing extra lines to the other in a load-leveling strategy. During the 1930s, Postal Telegraph was Western Union's "biggest cash customer," according to one WU manager—if a PT telegram was destined for a city where PT had no office, PT would handle it as far as possible and then bring it to WU for final transmission. Similarly, PT paid AT&T over $2 million per year for long-distance telephone service used to deliver messages to towns where PT had no facilities, in order to compete with WU in those places. And WU signed a contract with the New York City post office as early as 1880 to set up telegraph offices in post office locations. By the mid-1880s, New York City department stores like Wanamaker's, Macy's, and Field's offered all three communications services to customers in their stores at once.[4]

Yet the internetwork was not only a geography of shared wires and offices; it was also a geography of shared labor. This chapter considers how the telegraph messengers, first in concert with the post office, and later together with the telephone company, worked daily to tie their own telegraph network into the wider internetwork, such that the whole became more than the sum of its parts.

The story of the telegraph in America is an exception compared to that of the rest of the world. As mentioned in chapter 1, when offered the chance by Samuel Morse in 1844 to purchase the first prototype telegraph line erected between Baltimore and Washington, Congress declined (despite the objections of the postmaster general), thereby opening the telegraph network to private capital. This was not the only way to run a telegraph industry. Only two years after Western Union merged with its rivals in 1866 to become a "natural monopoly," England passed the Telegraph Act of 1868, which "authorised the Post Office to purchase the telegraph companies and the telegraphic business of the railway companies," and mandated that service be extended to unprofitable areas. Arguments for nationalization of the telegraph in the United States would persist through the 1880s and 1890s, especially after the Jay Gould takeover scandals, but Western Union was always successful in fending them off. Even the short-lived government takeover of the telegraph in World War I, something the postmaster general had been pushing for since at least 1913, had little effect on the industry (as described in chapter 9).[5]

Part of Western Union's strategy of resisting post office takeover hinged on the messengers. As early as the 1850s, even though the young and sparse telegraph system made regular use of the post office to handle messages where lines were down or where lines had not yet been strung, so-called final delivery of a telegram had to be made by a company representative—a messenger boy—if at all possible. The main factor in this choice was likely the cost of delivery, with telegraph companies paying messengers a penny or two per message and advertising "delivery at no extra charge" to their customers. Messengers were expected to speed telegrams directly to recipients, instead of following a letter carrier's daily schedule. But the use of messengers also helped to distance the privately owned telegraph from the publicly owned post office.

After all, the telegraph might easily have utilized post office labor instead of messenger labor, at least in some cities. Privately employed letter carriers began delivering the U.S. mail to New York City residences as early as 1855, for only 1¢ to 2¢ a letter—the same as the messenger boy wage. In 1863, a decade before ADT re-engineered the messenger force, the post office introduced free delivery of mail in the 49 largest U.S. cities, employing 449 letter carriers nationwide (New York City had 137 carriers alone, the largest concentration in the United States). Prices on "drop letters" (letters sent and received within the same city) were raised to pay for the free delivery of all mail, but those local letters still only cost 2¢ each—again, the same as a messenger boy's wage.[6]

Had the telegraphs used these postal workers for delivery in the 1860s, speed might have suffered a bit, but letter carriers could have funnelled new messages back into the telegraph system from their daily pick-ups, a useful trade-off. As it was, it took until the 1872 ADT call-box system for messengers to be summoned when a telegram had to be sent. Even then, WU president William Orton found himself arguing against a government takeover of the telegraphs by inflating the cost of messenger delivery, claiming, "It has cost the Western Union Company, for several years, an average of 2½ cents per message for delivery. In the large cities the cost ranges from 6 to 10 cents. At country stations the cost of delivery is so great that a special charge is made to cover the extra cost." But Orton's "6 to 10 cents" estimate for "large cities" was in direct contradiction to the contracts his company had just recently secured with ADT. Instead, Orton's inflated "delivery cost" actually included "the clerical labor of receiving messages, keeping the accounts, making out returns, providing blank forms and other stationery, furnishing stamps, and receiving and paying over the proceeds, providing office room, warm and lighted." Focusing only on "delivery" differences with the post office sent an important message: only a privately held employer of child labor, not a publicly funded employer of adult labor, could efficiently deliver telegrams on demand.[7]

Such arguments succeeded, and the post office and the telegraphs remained separate institutions in the United States, a difference that contributed to their being seen as competitors, even though up until 1910 they offered entirely different information services in terms of message speed, spatial coverage, and price. As described in chapter 2, the original invention of the "telegram" created a product that was much more expensive than the post office letter rate, but that traveled much more quickly. The subsequent creation of telegram day letters and night letters, described in chapter 5, carved out an in-between niche for the telegraph companies to sell a slower information product at a slightly discounted price, while load-leveling their wire plant and labor force at the same time. As long as the post office continued to operate at "railroad speed," even overnight telegrams would speed past the surface mail.[8]

In the early part of the twentieth century, it was the telegraph messenger service that was encroaching on the post office. Messengers were often contracted to deliver "samples, periodicals, letters, advertising literature, catalogues," and other items that were otherwise sent by second-class post. In 1903, WU messengers could blanket a neighborhood with a one-pound publication for the same 1¢ as the post office second-class rate, but returned a signed receipt with each delivery (to be used by the sender of the magazine in assembling mailing lists of potential consumers).[9]

In the 1920s, airmail threatened to reverse the competitive relationship between the telegraph and the post office. With the end of World War I, the first permanent daytime U.S. airmail route began in 1919, linking Chicago to New York via Cleveland. By 1923, airmail service could transport a letter from New York City to San Francisco in thirty-two hours (train service took one hundred hours). The service expanded quickly. In 1924, permanent continuous airmail service began—meaning planes flew at night as well as during the day. But the real boost to the service came a year later, when the Kelly Law allowed the post office to hire private airlines to carry the mail. By 1928, private airlines were handling all airmail traffic, and airmail rates were cut from 10¢ to 5¢—a rate cut that, Western Union argued, represented a significant (and unfair) government subsidy.[10]

At first, WU responded to the threat of airmail by touting the telegram's psychological advantage over the letter. In the very first issue of their public relations sheet *Dots and Dashes*, they argued: "The telegram is given priority over other forms of communication and when a business man arrives at his desk in the morning, he will generally find his telegrams lying on top of his mail." As discussed in chapter 5, key to this psychological importance of telegrams was the telegraph messenger: not only did his servile, uniformed presence imply the importance of the communication, but his movement outside of the mails and into the corporate office was what physically allowed the businessman's telegram to appear "on top of his mail."[11]

Later, WU shifted its advertising strategy to "scientifically" tallying up the "true" cost of sending a letter. The company's calculations were probably inspired by William Leffingwell's 1926 *Office Appliance Manual*, a classic of scientific office management that warned, "Investigation reveals the fact that a large percentage of the day's labor in the average office is expended in the dispatch of mail." For example, in 1927 Western Union reasoned that a business-man making $3,000 per year who "dictates ten letters per hour" to a stenographer making $1,500 per year was actually paying nearly 30¢ per letter. Further, "If the salary of the dictator is $6,000 a year the cost of the letter becomes 42¢. If his salary is $10,000 the cost of the letter becomes 57¢, and if it is $15,000 it mounts to 77¢." Western Union asked its customers: Was a 60¢ telegram really that expensive? What about an intracity telegram costing 24¢, or a coast-to-coast telegram for $1.20? The implication was that the more highly paid the executive was, and the greater the distance of communication, the more cost effective a telegram became. Here again, the activity of the messenger contributed to this selling point, as the messenger provided a very clear example of the idea that someone else was doing the work—not the sender, or the sender's secretary, or anyone employed to handle mail in the sender's organization. The messenger even brought fresh telegram blanks and pencils.[12]

Nevertheless, both the telegraph industry and its government regulators feared that airmail was taking a toll on telegram sales, since through the 1930s total revenue from telegraph traffic dropped by 23 percent. Western Union used the drop to request a 15 percent rate increase, but in 1938 the FCC denied the rate hike, arguing that raising telegraph rates would simply push more consumers to airmail. Postal Telegraph, trying at the time to convince Congress to authorize a merger with WU, agreed that airmail was a "cherry-picking" threat. According to the testimony of one PT trustee, "We have essentially the same mechanics for handling a 20-cent message as we have for handling a $2 message, and this development of air mail again hits at the most profitable section of our business." By the 1940s, the FCC commissioner agreed: "Fifteen or twenty years ago it took 5 days to get a letter from coast to coast. Today it takes 1."[13]

Of course, intercity airmail revenues were still only a fraction of intercity telegram revenues at this time, and as late as 1938 the rails still carried some 85 percent of all the mail anyway. But airmail threatened to become ever more popular, reversing the long-standing claim of the telegraph that it "annihilated" the speed of transportation through electrical communication. In 1941, Congress predicted that "this will remain a 'sick' industry until it has reduced costs through mechanization [such] that it can attract volume business through 'postalization' of rates and other improvements in service."[14]

But government-subsidized competition is only half of the story of the relationship between the telegraph and the post office. The telegraph companies neglected to advertise the many ways that they used the physical transportation network of the post office to help their own electrical communication network to function. Like any large national business of the time, they relied on the post office for certain command and control functions. Even though they themselves were in the communications business, a large part of the telegraph industry's own communication had to be mailed. As early as 1866, WU rules stated, "*The Official Correspondence* over the wires between Managers, Operators, and other employees of the Company, must be limited to matters of *an urgent nature and that will not bear the delay of the mail.—Use the mail* for all matters that will not suffer by the delay." The practice of using the mail for command and control continued as WU rationalized its interoffice communication into the *Journal of the Telegraph*, which reported "all changes of tariff, names of new offices opened, and offices closed," and all changes in company rules. Western Union apparently couldn't afford to tie up its own wires and operators with such communications.[15]

Telegraph companies also used the mail to plug gaps in their networks, especially in the late nineteenth century, when those networks were still being expanded, or when storms and floods knocked down lines. Especially in times of labor strife, such as during the failed Great Strike of 1883, WU offices that remained open mailed their telegrams using the post office. Customers paid directly for such postage, as the WU rules from 1870 explicitly stated: "Whenever a message is to be dropped in the post-office at its place of destination, the receiver will collect for postage and add it to the check."[16]

The steady improvement in postal service across the United States—without dramatic rises in the cost of stamps—encouraged the telegraphs to rely on the government for final delivery more and more frequently. After extending free city delivery to towns of over 10,000 residents in 1887 (provided postal revenues in those towns were at least $10,000 per year), the network could count 58,999 post offices across the nation, employing 150,000 people—8,257 of them working as city delivery mail carriers, in 401 towns and cities. In 1896, the post office began to experiment with rural free delivery (RFD), bringing the mail directly to and from farm homes. By 1902, RFD had closed nearly half of the 70,000 small, rural fourth-class post offices, but the spatial extent of the post office network actually increased. The post office was of course legally charged with the mission of serving the entire nation—and even though WU president Newcomb Carlton was fond of claiming that "there are two national systems of communication universal in character and open to the public at low cost—The United States Mail and the telegraph system," neither WU nor PT had any such legal obligation.[17]

Even as the telegraph network matured in the twentieth century, it still used the post office regularly when messenger delivery failed. As early as 1870, WU would send a notice through the mail when a message could not be delivered, holding on to the original telegram itself. But messengers were expected to contact the post office first to check the original address. In 1921 (and again in 1925), the assistant postmaster general issued an order stating, "Postmasters will hereafter comply as far as practicable with requests for street addresses to enable the messengers of telegraph companies to deliver messages." One former messenger from Long Island, New York, remembered, "we called [the] post office when addresses were obviously incorrect or for address changes. We mailed notices for undeliverable telegrams in the chance that the addressee would call us for a message being held."[18]

Just as the telegraph was using the post office to make up for messenger shortcomings, the post office was employing labor not unlike the messenger boys to help it better compete with the telegraph. In 1885, the post office began its Special Delivery service: for a fee, the post office would ensure that a letter was delivered immediately upon arrival at its final station, day or night, instead of waiting for the twice-a-day carrier route. Special Delivery was initially offered in any city with a population of four thousand or more.[19]

Right from the start, the relationship of Special Delivery mail to the telegraph messenger service was a complex one. In the 1880s, "Some district companies made private arrangements with postal authorities to let messengers convey Special Delivery letters," according to ADT's official history, a practice that continued into the twentieth century. Western Union advertised that its messengers carried the mail in Atlanta, Georgia, in 1916, and one messenger who worked in Deadwood, South Dakota, in 1933 recalled that he handled not only Special Delivery mail but Railway Express packages and "door-to-door deliveries of samples." On the other hand, by 1895 the *Telegraph Age* reported that Special Delivery service was making inroads on local telegraph traffic: "For 12 cents a person can have a letter or package delivered to distant points in a city or town . . . and the service is gradually taking the place of the telegraph for metropolitan business." A 1919 business text advised regular use of Special Delivery as well.[20]

No matter which network was benefiting the most from Special Delivery, there was a striking parallel between the new Special Delivery workers and the old telegraph messengers. Special Delivery mail messengers and substitute letter carriers who performed Special Delivery tasks were contingent, subcontracted, piece-wage workers just like the telegraph boys, even though other post office employees were part of the government civil service. In the early 1900s, substitute carriers were technically employed for a wage of $1 per year, which required them to show up at their local post office at least once in the morning (sometimes also at noon) to see if they were needed that day. If needed, they would be paid a day wage; if not, they were paid nothing. After three to five years of this, they could qualify for the civil service and become regular carriers. By the 1920s, post office messengers were legally supposed to be sixteen years old, slightly older than the fourteen-year-old telegraph boys; but during World War I, with the shortage of labor, the post office had permitted children aged fourteen and even younger to work as Special Delivery messengers. And just as with the increased employment of telegraph messengers in the 1920s, "a ruling by the Interstate Commerce Commission [that] reinterpreted the obligations of public carriers" resulted in an increase in the number of post office mail messengers as well.[21]

Messengers in both networks recognized the similarity in duties and the disparity in wages between their work and the work of regular letter carriers, and fought for equal treatment. In a 1938 hearing before the Senate Committee on Interstate Commerce, the American Communications Association (a telegraph union) argued on behalf of the telegraph messengers, saying that the 23,000 children working for WU and PT, who averaged $8 per week for forty-eight to sixty hours of work, performed work similar to adult post office carriers, "grown men who have passed a rigorous physical examination." The biggest difference between the post office Special Delivery messengers and the telegraph messengers was that the post office messengers had a viable career path. By 1936, a substitute letter carrier might make only 50¢ to 65¢ per hour (as compared to a telegraph messenger's 12¢ per hour), but after thirty-one full-time weeks, he could be promoted to a regular city carrier. City letter carriers had to be at least eighteen years old, and had to meet minimum height, weight, and strength requirements. Applicants took a civil service exam, which under 50 percent passed. By 1940, these letter carriers made $1,700 to $2,100 per year, a relatively low wage for the time, but still much higher than a messenger's $416 per year.[22]

Ironically, these same letter carriers used comparisons to messenger boys to argue for better wages and conditions themselves. As one of the first presidents of the National Association of Letter Carriers said, "We hold that [the letter carrier] is not a messenger boy, but that he is an intelligent part and parcel of the social and industrial organization of this land of ours." Yet as the intertwined history of the two labor forces reveals, both messengers and letter carriers were "intelligent parts and parcels" of the social and industrial information internetwork.[23]

Unlike the situation between the telegraph and the post office networks, the idea of cooperation between the telegraph and telephone networks always seemed to make immediate sense—after all, both networks basically sent electrical information over wires. Why not combine them into one system? Didn't the Bell company itself have the hubris to rename itself American Telegraph *and* Telephone? Indeed, at two key moments—at the birth of local telephony in the 1870s, and during the rebirth of long-distance telephony in the 1910s—it seemed that either WU or AT&T might very well end up controlling both networks. What often goes unrecognized about each of these historical moments is that the costs and benefits of combination depended not only on shared wires, but also on shared messenger labor.

One of the great ironies of U.S. communications history is that in 1875, before Alexander Graham Bell had even patented the telephone or started his own company, he demonstrated the device to WU president William Orton, who refused to purchase it. Explanations for this refusal differ, but a common argument is that the deal soured when Orton found out that his business enemy Gardiner Hubbard was involved (both as Bell's father-in-law and as Bell's financial backer). Hubbard had pressured the government in 1868 to nationalize the telegraph system and merge it with the post office, as had been done that year in England—the resulting bill nearly passed Congress in 1874. In any case, shunned by WU (as the story goes), Alexander Bell patented the telephone in 1876 and formed his own Bell Telephone Company a year later. But Orton died in 1878, passing control of Western Union to Norvin Green, and soon WU had changed its opinion of the telephone, seeing it as a potential competitive threat.[24]

Whatever its fears of a telephonic future might have been, Western Union was clearly troubled by the fact that Bell targeted district telegraph companies such as ADT as the logical sites for installing new telephone systems and switchboards. Even if subscribers couldn't talk to each other yet, there was value in allowing them to call the central office to describe their request, instead of simply buzzing a call-box and waiting for a messenger. Western Union was already using its newly allied district telegraph service as a venue for competition with Jay Gould's A&P telegraph (described in chapter 3); now WU had another reason to worry about the ADT contracts.[25]

The Bell threat to ADT led Western Union to assemble its own telephone company. Though WU had no claim in the Bell patents, it owned the patents of Elisha Gray (for the principle of the telephone), Amos Dolbear (for improvements), and Thomas Edison (for a better carbon-button, variable-resistance transmitter). Western Union created the American Speaking Telephone Company to hold these patents, distributing phones through its existing Gold and Stock Telegraph Company and hooking up exchanges through local ADT franchises.[26]

Bell soon brought a patent infringement suit against American Speaking Telephone, commonly called "the Dowd suit." But WU president Green was looking for a market-sharing agreement, and in November 1879 he got it. The terms of the agreement separated the telegraph and the telephone publicly, but linked the two networks behind the scenes: Bell would pay WU a 20 percent royalty on all telephone rentals over the seventeen-year tenure of the Bell patents (expiring by 1894), while WU gave Bell all of WU's telephone inventions plus the right to use WU pole lines. Western Union sold its telephone exchanges to Bell, and Bell agreed to turn over all phoned-in telegraph messages to WU for a 15 percent commission on such sales. Thus the two networks were to serve different purposes—telephone exchanges were limited in size to a fifteen-mile radius from a central office, to be used for "personal conversation" only, and not for "the transmission of general business messages, market quotations, or news for sale or publication in competition with the business of Western Union."[27]

As one telephone historian put it, "Just why the forty-million-dollar telegraph giant . . . dropped its competing patent claims to the telephone is one of the more curious and fascinating questions in the history of American business." Yet at the time, it was easier for contemporaries to see the two networks as fundamentally different. Even when the Bell patents expired in 1893, a telegraph expert could write, "With the present apparatus and mode of working, the telephone is not adapted for regular telegraphic business, and is not likely to supersede the ordinary apparatus used in telegraphy." The reason? "We can telegraph much faster than we can speak, and this is of more importance than speech; hence any device that retards speed can *never* take the place of the telegraph apparatus; besides this, is its liability to error, and it not being practically possible to secure the privacy which the telephones require."[28]

What is missing from this story of the initial relationship between the telegraph and the telephone is the participation of messenger boys, a fact that likely affected contemporary perceptions of the long-run feasibility of each service. The first workable telephone exchange was developed in New Haven, Connecticut, in 1878, allowing the switching of calls between twenty-one subscribers—with former messengers working the switchboards. Ironically, this exchange advertised itself as a replacement for messenger service, claiming, "your wife may order your dinner, a hack, your family physician, etc., all by Telephone without leaving the house or trusting servants or messengers to do it." But at the same time, messengers were crucial to this service as well, as the new exchange announced, "An efficient corps of neatly uniformed messengers will be employed to answer calls" (and to perform odd jobs at 25¢ per hour). After all, until the entire business community was on the same phone system, phoned orders from patrons would have to be relayed to actual merchants through messengers. The subscription price for this telephone service was set at $18 per year, even though that figure was likely to result in a loss, because that was the same fee that ADT charged for its messenger-based call-box service.[29]

Messenger boys seemed to relish operating the new telephone technology. One Boston boy operator in the 1870s described "A number of boys, wearing carpet slippers to deaden the noise of their feet, running up and down before a switchboard twenty feet long, three or four of them sometimes making a jump to answer the same drop that had fallen and yelling their lungs out through a hand telephone." Bell telephone engineer John Carty, who at age nineteen worked in a telephone exchange in 1879 earning $5 per week, remembered that his fellow operators "were not old enough to be talked to like men, and they were not young enough to be spanked [like] children." But no matter what their age, just as in telegraph offices, boys who operated the early district telephone exchanges did more than simply route calls and carry messages. Subscribers were few, meaning call volume was small, so boys did janatorial work, collected bills, and even helped to repair lines.[30]

As call volume increased in the 1880s, messenger boys, thought to be too rough on both the customers and the equipment, began to be replaced by girls hired specifically as telephone operators. One boy operator recalled, "New boys were initiated by shocks applied through a number of batteries and were also expected to put on boxing gloves to prove their prowess." Historian Stephen Norwood notes, "The companies objected to the boys' boisterous behavior, inattention to instructions, and insolence toward subscribers, including the use of profane language. They assumed that young women and girls were capable of a greater civility and could more easily tolerate monotonous work and low wages." And one contemporary remembered, "so many of these boys did 'sass' back that telephone exchanges became, in many places, exchanges of loud and lurid language—to such an extent that discipline was impossible and boys had to be fired almost faster than they were hired."[31]

The women who replaced the boys as operators were paid between $6.50 and $7 per week in the 1880s and 1890s, as opposed to messenger wages of $4.50 per week. The telephone companies sought not just any women, but women with particular characteristics: they were to be unmarried, English-speaking girls aged seventeen to twenty-six, with two years of high school education. Managers thought that educated girls would interact better with the affluent telephone customers. By 1894, a *Telegraph Age* cartoon could depict a messenger boy working a switchboard as an unusual, comical occurrence. Apparently, however, boys still sometimes covered the night switchboard duties, even as late as 1904.[32]

The relationship between telegraph call-box systems, local telephone exchanges, and messenger boys evolved as telephone technology improved. In 1900, ADT in St. Paul, Minneapolis, replaced its old address-printing call-boxes with "autophone" call-boxes, hybrid units that not only announced the ID number of the buzzing customer, but allowed for voice transmission of the customer's request (see figure 7.1). Autophones did not allow subscribers to talk to each other, but were instead intended to save time and effort by removing the need for a messenger to come to the customer to find out what the customer wanted. However, messenger service itself was still intact, because if the customer had a pickup, the messenger would still have to come get it. And even with autophones or other tricks for sending a "return signal" back to the cus-

Figure 7.1. *The autophone, 1910*

Source: *T Age* (June 16, 1900), 248.

tomer's call-box, many systems remained on standard one-way call-boxes for labor reasons—after all, if customers expected an immediate answer to their call-box requests, a clerk fixed in place at the district office would have to take the time to actually provide that answer (rather than, say, waiting for several call-box requests to be recorded on punched tape before sending out a single boy to handle them all).[33]

In the late-1870s maneuverings of WU and Bell to plan the future of the telecommunications market over the next twenty years, messengers may not have been of the greatest concern to executives and managers. But just as in the Jay Gould and Postal Telegraph battles described in chapter 3, focusing on the messenger connection shows how competition for wires and customers actually played out on the ground, in the sites where telephone service depended on the ability of a boy to switch circuits, to speak pleasantly, and to run the occasional print message—activities that weren't much different from those in the telegraph network. Even after the 1879 patent deal, telephone exchanges relied on messengers to help grow their businesses, as a pamphlet instructing new users of the "Chicago Telephonic Exchange" illustrates. After four paragraphs of detailed instructions in how to use the phone, in small type, the pamphlet ended with this notice in large type: "MESSENGERS Furnished Promptly by notifying Central Office." With standard practices of business-to-business and business-to-consumer communication so tied to messengers, the use of boys helped make it difficult for others to see how a messengerless communication system could ever threaten the dominance of the telegraph.[34]

Through the rest of the nineteenth century, before the telephone could compete with the telegraph for intercity service, the two networks coexisted in relative harmony. But once long-distance telephony became feasible after the turn of the century, the main product difference that had allowed the truce between WU and Bell was gone. In 1901, in a gesture showing the importance of long-distance traffic to their future business, American Telephone and Telegraph, the Bell subsidiary started in 1885 to manage long-distance efforts, changed its position to become the parent company of the rest of the Bell system.[35]

By 1907, AT&T had lost half of its monopoly phone market to new companies appearing after the original Bell patents expired in the 1890s. AT&T tried to expand in the face of this competition, financed through heavy borrowing and the sale of stock to financier J. P. Morgan. But banks would no longer allow AT&T to acquire further bonds to finance its growth. Thus New York investors led by Morgan ousted AT&T's former Boston investors, and brought in Theodore Vail as AT&T president—even moving the corporate headquarters and the engineering department out of Boston and down to New York City. With telephone stock declining, Vail risked issuing 220,000 shares of new stock, helping AT&T weather the 1907 panic. He also fired twelve thousand Western Electric manufacturing workers and reorganized the research and development department. By 1908, AT&T had begun to advertise its competitive advantage as "One Policy, One System, Universal Service"—publicly announcing its intention to eliminate duplicate telephone providers.[36]

But "Universal Service" meant addressing the nascent "Telgraph" portion of American Telephone and Telegraph as well. Late in 1910, after first flirting with the idea of purchasing Postal Telegraph to oppose Western Union, AT&T bought 300,000 shares of WU stock (or 30 percent) from the estate of former WU baron Jay Gould to assume control of the telegraph giant. Vail became president of WU as well as of AT&T, current WU president Robert Clowry moved to the executive board, and Gould's man Thomas Eckert was removed from the executive board altogether—clear signs that an older era of telegraphy had ended.[37]

Certainly the notion of a monopoly in electrical communications was attractive for many reasons. For example, by 1912 AT&T had placed 25,000 new model "multi-coin collector" pay phones in Manhattan and the Bronx, phones that patrons could use to call Western Union free of charge to dictate a telegram. But receiving and delivering telegrams over the phone had long been a part of individual WU contracts with local telephone networks. As early as 1881, WU executives decided to pay a maximum of 2¢ per message to phone companies for such "delivery" services, with the stipulation that "This Company not to be responsible for errors made in deliveries through the Telephone."[38]

Vail's goal for interconnection was actually broader than just telephoning telegrams. He wrote, "The defects of the telegraph system of today are collection and delivery; old-fashioned, primitive, slow, unreliable, and expensive. It costs more to collect or deliver a message than it does to transmit it. It is like riding from Boston to New York over 220 miles and then paying 20% more to get yourself and your gripsack down to your hotel." Vail wanted to combine all telegraph and telephone facilities, both lines and offices, "to eliminate the imperfections in collection and delivery and also by joint occupancy and joint use with telephone companies get all offices in the custody and charge of employees who have only one interest to serve." As Vail put it in the AT&T annual report for 1911, "This is what the Bell System aims to be—one system with common policy, common purpose and common action; comprehensive, universal, interdependent, intercommunicating; like the highway system of the country, extending from every door to every other door; affording *electrical communication of every kind*, from every one at every place to every one at every other place"[39] (see figure 7.2).

But to contemporaries, the inefficiencies of the telegraph were not just represented by its many offices or its overlapping wires; inefficiency was embodied in the messenger boy. One early historian of the telephone criticized Western Union in 1910 for "employing as large a force of messenger-boys as the army that marched with General Sherman from Atlanta to the sea," and predicted that the greatest benefit of telephone-telegraph combination would be "in removing the trudging little messenger-boy from the streets and sending him either to school or to learn some useful trade." The trudging boy may have been slow, but at the same time he was certainly inexpensive. The New York state legislature feared that Vail's motivation for replacing the messenger boy with the telephone was simply to sign up more telephone subscribers—customers who would soon pay more to AT&T in phone bills than they ever paid Western Union in messenger charges.[40]

Vail quickly began to merge the telegraph and telephone wire plants. Long-distance transmission of both products could now be handled over lines of either system simultaneously, due to what was called the "phantom circuit" technique: "When two telephone circuits were used in connection with each other, three telephone conversations could be carried on and eight telegraph messages sent at the same time." But as described in chapter 5, Vail relied less on transforming the messenger than on transforming the message. He introduced the day letter and the night letter, telegrams that would cost less to the public and travel at a slower speed, but that would be transmitted during the network's slow hours, taking advantage of fixed capital and overhead investments.[41]

Figure 7.2. *AT&T's vision of a combined Bell and Western Union, 1911*

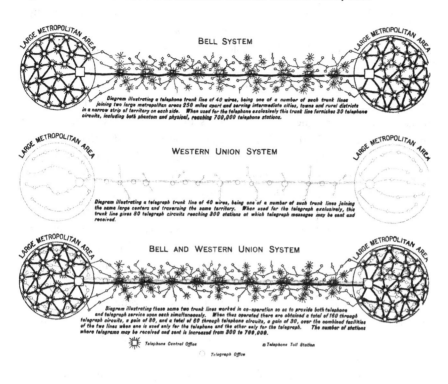

BELL SYSTEM

Diagram illustrating a telephone trunk line of 40 wires, being one of a number of such trunk lines joining two large metropolitan areas 250 miles apart and serving intermediate cities, towns and rural districts in a narrow strip of territory on each side. When used for the telephone exclusively this trunk line furnishes 30 telephone circuits, including both phantom and physical, reaching 700,000 telephone stations.

WESTERN UNION SYSTEM

Diagram illustrating a telegraph trunk line of 40 wires, being one of a number of such trunk lines joining the same large centers and traversing the same territory. When used for the telegraph exclusively, this trunk line gives 80 telegraph circuits reaching 300 stations at which telegraph messages may be sent and received.

BELL AND WESTERN UNION SYSTEM

Diagram illustrating these same two trunk lines worked in co-operation so as to provide both telephone and telegraph service upon each simultaneously. When thus operated there are obtained a total of 160 through telegraph circuits, a gain of 80, and a total of 60 through telephone circuits, a gain of 30, over the combined facilities of the two lines when one is used only for the telephone and the other only for the telegraph. The number of stations where telegrams may be received and sent is increased from 300 to 700,000.

Telephone Central Office   □ Telephone Toll Station

◌ Telegraph Office

Source: AT&T, *Annual Report* for 1911 (New York: AT&T, 1912).

More efficient or not, Vail's AT&T/WU giant was not to last. In 1911, the Supreme Court, under antitrust law, broke up the Standard Oil trust and the American Tobacco trust. Vail tried to explain in a 1913 *Atlantic Monthly* article just why AT&T's case was different, calling the telephone and telegraph company a "public utility" (even though it was privately owned) and praising its monopoly status as "survival of the strongest, if not the fittest," the result of "economic and natural laws." But Vail's public utility was not to be. That same year, the attorney general asked the Interstate Commerce Commission to investigate AT&T and WU for several monopoly charges: (1) that they were buying up smaller, competing phone companies; (2) that they refused to interconnect with competing phone companies, especially in long-distance lines; and (3) that they discriminated against competing telegraph companies like Postal Telegraph by automatically funneling phoned-in telegram requests to WU. Coupled with this investigation was the fact that in 1913 J. P. Morgan died leaving AT&T without a strong advocate for defending its monopoly. AT&T relented rather than be "trustbusted," agreeing to what is known as the "Kingsbury

Commitment" (after the AT&T vice president Nathan Kingsbury, who made the proposal in a letter to the U.S. attorney general): (1) they would stop buying up smaller competitors; (2) they would allow interconnections between competing systems; and (3) they would divest from WU.[42]

The Kingsbury Commitment did not sever all ties between AT&T and WU, however. AT&T did dispose of its WU stock, at a loss, and Vail stepped down from the presidency of WU in 1914; but the two companies continued to make individual agreements echoing Vail's ideas, like sharing long-distance trunk lines. In 1915, the Postal Telegraph journal asked, "Does the Bell Telephone control the Western Union?" and concluded, "It rather looks that way" because in Bell phone booths, signs still directed customers to "say 'Western Union'" into the telephone receiver in order to send a telegram, and because calling a Bell telephone operator and asking for PT often elicited a response saying that AT&T only did telegram business with WU. Sure enough, as late as 1926, WU would claim that "virtually every coin box telephone is a telegraph station." A customer who called the telegraph office from a pay phone, paying for the call and then dictating the telegram, could then drop coins in the phone to pay for the message immediately. Similarly, from a home or business phone a customer could call WU and dictate a telegram, and the cost would appear on that customer's next phone bill. AT&T took a 5 percent commission on each of these telegrams.[43]

Even without the public phone booths, the telegraph industry's use of the telephone, spurred on by the brief merger, continued with new force in the 1920s. With AT&T trying to avoid being seen as tied to WU, both PT and WU were able to lease lines from AT&T in order to send telegraph traffic over these phone lines. Some of this traffic apparently used the "carrier-current" method of adding up to ten duplex telegraph circuits on a single pair of telephone wires, using frequencies well above the range of the human voice for the telegraph traffic—and allowing the telelgraph companies to defer new construction on their own lines. Such practices would continue into the 1950s, leading one writer in *Fortune* magazine to comment at midcentury that WU was "crucially dependent" on AT&T for at least 60 percent of WU's own circuits.[44]

Again, not only circuits, but people were involved in this practice. As early as the 1920s, both WU and PT had set up "telephone delivery departments" in large cities (see figure 7.3). These departments were staffed by, and often supervised by, women. Thus they represented another aspect of the feminization and mechanization of telegraphy in the early twentieth century. The goal of these telephone departments was to eliminate the messenger boy. In 1921, the female supervisor of the Chicago Postal Telegraph telephone department talked of still having to "educate the public" in this new practice: "While there is still a pref-

Figure 7.3. *Western Union telephone department, New York City, 1910*

Source: *T&T Age* (November 16, 1910) 764–65.

erence for delivery and pick-up of telegrams by messenger, it has proved impractical in certain sparsely settled business districts, far removed from a branch office. In order to minimize delays, we have been forced to resort to the telephone."[45]

At Postal Telegraph, the work of the telephone department was divided into specific tasks. Some women only "delivered" telegrams by phone all day, others only "collected" telegrams that were phoned in by customers. Some women operated on the local switchboard, others on the long-distance one. And some only furnished rates and answered questions for the public. In WU, the division of labor was similar: public operators received and delivered telegrams over the phone, and "private branch exchange operators" connected individual *telegraph* customers to each other over leased *telephone* lines. Some operators were even used, in a load-leveling strategy, for internal communication between busy offices, with normal telegrams actually telephoned between branch locations during the busiest times of day. Western Union "branch operators" relayed telegrams over the phone between intracity offices, between city and way-station offices, and between major cities along trunk lines. Thus telegrams moving through such telephone operators had been typed in once already, were dictated over the phone, and later were retyped once again—all before being actually sent on the key.[46]

This example highlights an important tension: the telephone departments were meant to speed message transmission at lower costs by eliminating the need to physically render the telegrams in a form that messengers could carry, but the process was never that simple, and there were complex connections between information in its virtual and physical forms. For example, to "collect" a phone telegram at PT, a telephone operator listened to the message as it was dictated over the phone by the customer, typing it into a "noiseless typewriter." From there, she sent the printed page via pneumatic tube to the operating room, where it had to be retyped again into the automatic telegraph system! Finally, of course, the telegram was rendered in print once more at the receiving end. Delivering a printed telegram over the phone was also a complicated process. "Senders," as the women were called, had to read well, speak distinctly, and understand business terms. Instead of reading the printed telegram in order, they gave the return address first, then the signature, then the body of the message. After two busy signals (a situation called a "can't raise"), the telegram would be sent by messenger. If "collect" messages were delivered by phone, a bill would be sent to the receiver through the mail, and at the end of each day the print copies of all messages that had been telephone-delivered were mailed anyway—meaning that both the post office *and* the telephone company were subsidized when the telegram traveled without messenger delivery.[47]

Even with its convoluted translation of telegrams from physical to virtual and back again, the collection and delivery of telegrams over the phone threatened to eliminate the telegraph messengers. Whenever messengers went on strike, managers confidently predicted that service would remain uninterrupted: "We are more than able to make up for the loss of messengers by using the telephone for delivering messages and if the boys don't look out they will find that the telephone will take their places in many cases." But in reality, all through the 1920s in New York City, only 300,000 of the 9.5 million telegrams delivered each month were delivered by telephone, or around 3 percent. Fifteen years after Vail's dream, even without AT&T control over the industry, the telephone and telegraph networks *were* interconnected—but messengers still persisted.[48]

The persistence of the messenger turned out to be a blessing in disguise for the telephone industry, for messengers helped answer the question: How does one call somebody who doesn't have a phone? In the early days of telephony, the very advantage of synchronous, point-to-point, unmediated communications—the ability of one individual to talk to another directly if both were available at the same time—was also a liability. The telephone network did grow rapidly in the twentieth century, from 1.3 million telephones in 1900 to 19.6 million in 1931. But until *everyone* had a phone, subscribers could only call other subscribers.

This situation was exacerbated by the fact that before Bell agreed to interconnect competing local phone companies in 1914, patrons of one system had no way of calling patrons of another.[49]

Messengers were used as a solution to such problems as early as 1895: "The telephone companies make no charge for the service if the party at a distant point is not in or cannot be found. It makes no difference whether or not the party has a telephone, as in all probability he can be reached by messenger service, for which a nominal charge is made. If the party cannot be found an appointment can be made, for which no charge will be demanded until the second connection is secured." It is unclear whether the messengers in question were employees of the telephone companies or whether they were hired as needed from the local district telegraph companies; probably both methods were used.[50]

This practice persisted at least into the 1930s, when due to the Depression a cash-poor telephone subscriber might decide to cancel service, preferring, say, to pay for the upkeep of an automobile. A 1932 high school business text noted that "If the person you wish to call in a distant city has no telephone, you may arrange with the long-distance operator to reach him by means of a *messenger call*. This means that a messenger will be sent to the home or office of the person called to advise him that a telephone call awaits him at a nearby station." According to one old-time telegrapher, such messages might even be printed out as telegrams so there would be no mistake: "PLEASE GO TO A NEARBY TELEPHONE AND ASK FOR OPERATOR #—A PARTY WISHES TO SPEAK WITH YOU BY TELEPHONE." Thus at the same time that the telegraph industry was trying to replace its messengers with telephones, the telephone industry relied on messengers to relay messages from subscribers with phones to nonsubscribers without phones. A visit from a uniformed messenger remained the ultimate "person-to-person" call.[51]

This chapter has shown how the telephone and post office networks were not only competitors of the telegraph network, but partners with it as well. At different times and at different places, any one of these three networks might hold the master locational address information for homes and businesses in a city; at different times and places, one network might go down due to bad weather, labor strikes, or technological failure, and the other two networks would be used to cover the message flow; and even under regular operating circumstances, all three of the networks relied on each other for their day-to-day functioning, whether coordinating personnel movements by phone or handling billing through the mail.

In such an environment, to speak merely of "competition" or "cooperation" between these networks is not enough. The metaphor of "connection" is much more suggestive and generative, helping to reveal the physical and electrical internetwork of telephone, telegraph, and post office networks—an internetwork that was more than just the sum of its parts. In a consumer landscape where many lacked telephones or even a stable home or work address, the combination of these networks resulted in a sort of multimodal message delivery that functioned more reliably than any one network could on its own. Telegrams that did not need to be hand-delivered could be phoned; phone calls that could not be connected could be transferred to print messages; messages that could not be delivered by messenger could be mailed. There were even new services, unexpected "generative" bonuses, that came out of the combined system, such as the singing telegram, which started out as an option for phone delivery only, but eventually became part of the physical repertoire of the messenger boy (as discussed in chapter 5).

Telegraph messengers—and their counterparts in the post office and telephone networks—were crucial in connecting this analog internetwork; they enabled the emergent pattern of multimodal message delivery and dealt with all the problems that thwarted that delivery. From the point of view of the consumer, a message was merely placed in the hands of the messenger—the internetwork boundary-worker—and somehow it reached its destination.

CHAPTER EIGHT

THE MYTHS OF EDUCATION

He was now twenty-one years of age, he had changed jobs four times, had been idle between jobs and had begun to realize that he could have but little hope of getting a firm hold and a climbing grip upon the industrial ladder unless he knew a trade, or unless he be trained for some definite work.

—tale of a messenger from *The Boy and His Vocation*, 1925[1]

MANAGER: *Howard, if you want to stay with us, you'll have to learn to do things our way. You want a chance at advancement, don't you? Someday you might even become president of Western Union. How would you like that?*

MESSENGER: *The disadvantage of being president of the company is that you have no chance for advancement.*

—joke told by a Western Union manager, 1950.[2]

Early on in telegraphy, messengers working in low-traffic offices had a real chance to learn how to send and receive Morse code, a career potential that legitimized their long hours and low wages. But as described in chapter 3, when ADT appeared in 1872 the tenuous apprentice relationship between messengers and operators was severed, at least in large cities. Urban telegraph offices, where messengers became subcontracted piece workers, diverged from rural telegraph offices, where messengers were still directly hired as telegraph employees. Since a local ADT franchise was not a national telegraph company and was unable to offer a career path for the many messengers it hired, ADT was saddled with a nationwide problem of rampant messenger turnover. For the messengers, what had been an "internal" labor market, with rules and customs for eventual advancement into a booming high-tech field, had been replaced with an "external" labor market, where the only rule that applied was worker competition for a subsistence wage.[3]

When publicly questioned about their high turnover rates, managers relied on two strategies to explain frequent firings and resignations. First, they claimed that those who were fired didn't deserve a good job. In 1875, the *Telegrapher* reported, "[ADT's] system of supervision is so complete and effective, that the dishonest, indolent and incapable are quickly weeded out," leaving only the "more active, intelligent, gentlemanly and generally well-behaved community of boys" on the payroll. The second strategy, used to explain why so many messengers quit voluntarily, invoked some form of what could be called the "messenger career advancement myth," claiming that boys found better jobs thanks to their messenger service. Worthy messengers would naturally rise in the business world, in Horatio Alger "pluck and luck" fashion. The negative job attribute of high turnover was recast as a positive "placement office" function, in order to legitimize it.[4]

An early statement of this career myth was an 1870 Western Union editorial arguing, "A large number of telegraph messengers have risen to comparative opulence and respect. They rub against the active men of the world who are quick to discern merit. They get a chance to peer into the ways of mercantile life, and are touched with the desire to rise." The myth had two variations: rising in telegraphy, and rising in the general business world. Western Union president Newcomb Carlton invoked both versions in testimony before the 1915 Commission on Industrial Relations, saying, "We attempt to get a class of boys who can either develop into the service, or will make good office boys for somebody else." As time went on, first in cities and then in rural areas, the "external" variation of the advancement myth supplanted the "internal." It was this idea that messenger work was a "prep school to the world of business"—drawn from the economic imperative to continue to employ young boys in the face of child-labor challenges—that led the nation's largest telegraph company into the education business itself.[5]

The internal messenger advancement myth was rooted in the former messenger apprentice relationship. Western Union argued in 1875, "The duties of the messenger service eminently qualify for a comprehension of all other telegraphic labor," such that "after a fair service as messengers, must come, after competent training, the future operator, and manager, and superintendent." This argument might have been persuasive with official WU messengers, but it had less force as messenger work shifted to subcontracted ADT messengers. As the *Telegrapher* revealed, most ADT boys stayed for an average of only six months. In its first three years of business, ADT New York City had hired about eleven hundred boys per year to fill their five hundred slots—ten times the number of promotable positions in ADT.[6]

As the telegraph industry expanded and aged, the myth of internal advancement took on new force: just look at all the telegraph executives who started out in the "cradle" of the messenger department, managers argued. No doubt many truly believed the myth they perpetuated, because of their personal experiences. Belvedere Brooks, who became WU general manager in 1910, had started as a messenger at age twelve in 1871. Robert C. Clowry, WU president from 1902 to 1910, began as a messenger in 1852, supposedly working for free for six months in order to learn telegraphy. And WU messenger manager W. S. Fowler, who in the 1920s lamented about "what an everlasting job it is to keep our force of twelve thousand messenger boys adequately 'manned,'" was a messenger in 1903.[7]

But contrary to the myth, managerial positions were rarely filled by messengers who worked their way up the ranks; rather, future managers were picked through other means and, if needed, given "messenger experience" on the streets for a few weeks or months. In 1917, WU sought employees with high school and college education (rather than the grammar school education usually held by messengers) to train as office managers. These educated young men started as messengers temporarily to learn the business from the ground up, but soon moved on.[8]

This was the experience of one of WU's most famous managers, writer Henry Miller. In 1920, before embarking on his first novel, a dejected Miller tried to find work as a New York City messenger. He was inexplicably rejected for the job, but he pushed his case before a higher WU manager until his spunk earned him a managerial slot of his own. First, however, he himself, at age twenty-nine, would work as a "mutt" for a month or so, earning only $17 per week, in order to learn firsthand what messenger work was like. After the probationary period, Miller was promoted to a $60-per-week adult position, where he oversaw the yearly processing of ten thousand messengers to fill a constant force of about one thousand—hiring hundreds of new boys every month.[9]

Not only manager slots, but even clerk positions were increasingly closed to messengers. The variable nature of the telegraph meant that operator needs might fluctuate not only seasonally, but daily and even hourly; therefore, it made sense to fill clerk positions with employees who could act as operators as well. Opportunities to learn operator skills still existed for WU messengers in some places—in 1917, the Harrisburg, Pennsylvania office enrolled twenty of its bike messengers in telegraph school, "a clean-cut, wide-awake bunch." But in larger cities, Morse training no longer carried the weight it once had. A 1914 Chicago Board of Education investigation found that even at a company that offered one hour a day of operator training to its messengers, "Out of 337 boys employed . . . only 25 attend the school, since they are not paid for the time spent in training."[10]

Even messengers could see that the job of telegraph operator (like the parallel job of telephone operator) had become "feminized," meaning not only that young women were now the desired employees, but that the job was reconstructed as one appropriate for the (assumed) needs and abilities of women: wages were low, the career path was short, and the work was increasingly deskilled and mechanized. As early as 1859, WU had partnered with the Cooper Union in New York City to offer free telegraphy instruction to women. Especially after the failed 1907 operators' strike, the company began replacing older male Morse operators with younger female Automatic operators wherever feasible—a fact they were widely advertising by World War I. A young male messenger's chances of becoming an operator were reduced considerably.[11]

In a subtle acknowledgment of the shrinking opportunities for messengers as operators, managers increasingly claimed that messenger work would lead worthy boys to a career not in telegraphy but in business. The canonical example of this external advancement myth was the story of Andrew Carnegie, who, as WU frequently pointed out, started as a messenger himself in 1851 at the age of fifteen, earning only $2.50 per week. Carnegie later repeated the myth in his 1920 autobiography: "I do not know a situation in which a boy is more apt to attract attention, which is all a really clever boy requires in order to rise."[12]

ADT first used the external advancement myth when, only two years old, it was threatened by a New York State mandatory schooling law passed in 1874. The law directed that all New York children aged eight to fourteen attend school, a big problem for a company like ADT, which employed some four hundred New York City messengers, "of whom probably nine tenths are within the age affected by the law." Refuting charges that children belonged in school and not at work, ADT vice president E. B. Grant argued, "Our messengers, in the performance of their duties, are brought in contact with the leading business men of the city," and thus "more than a hundred have, within the past two years, found permanent places at an advance upon the wages we pay, with subscribers whom they have pleased."[13]

One practice supporting this external advancement myth was the hourly rental of ADT messengers for any kind of job a customer desired. During an 1880 messengers' strike, the *New York Times* reported, "Some of the striking messengers have secured places as office boys in brokers' offices. It is a common practice for boys to get such positions, the training they get in the District Telegraph Company's service being just what a broker's messenger needs." Yet a brokerage house had no incentive to hire full-time boys from the messenger force, since the reason a business contracted for boys in the first place was to hire labor only when needed, avoiding the expense of a full-time worker. More likely, the only alternative jobs available to messengers were messenger jobs with competing telegraph companies. In 1881, as the WU-allied ADT Philadelphia found itself competing with Jay Gould's upstart American Union (as described in chapter 2), the district company found itself "somewhat crippled by the opposition telegraph companies' inducing our messengers to leave our service and paying increased wages." Thus were the meager "advancement" opportunities of the messenger force.[14]

In contrast to the messenger advancement myth was the accusation by turn-of-the-century child-labor and education activists that messenger work was a blind-alley or dead-end job: it taught no skills, it led to no promotions (inside or outside), and it robbed children of precious time that should have been spent either in school or in learning an actual trade. Western Union and ADT would soon become targets of this Progressive campaign, eventually forcing the companies to act in order to preserve their young workforce.[15]

Child-labor laws were not new—as early as 1842, Massachusetts and Connecticut had laws limiting the working day to ten hours for children under age twelve. But for most of the nineteenth century, instead of trying to eliminate child labor outright (which would have been opposed both by poor working families and by firms that relied on the labor of children), reformers concentrated on nibbling away at what they saw as the excesses of child labor. Dangerous tasks, "immoral" locations, overexertion, and long work hours were targeted, but neither the low wages nor the right of employers to hire children were challenged.[16]

The idea of a dead-end job changed the terms of the debate: the rationale for the child taking a job was no longer simply to earn money, but to learn the skills necessary for a career in an urbanizing and industrializing world, a world where apprenticeship was a thing of the past. Just as reformers had studied which jobs were most dangerous to children, they now began to study which ones had the fewest career prospects. Susan Kingsbury's Massachusetts report of 1906 singled out department store work and errand work as canonical dead-end jobs: "[They] do not afford a living wage, and offer no opportunity for advancement to one. They are distinctly bad in influence, since the younger employee is so shifting, resulting in instability of character. When the child has reached 16 or 17, he or she must begin again at the bottom."[17]

As described in chapter 6, children's "street trading" was one of the "dangerous" jobs targeted in the early 1900s. Street traders like messengers were often cast as little entrepreneurs: "The messenger boy is a business man on his own—just as much so as the reporter working on space, the salesman on commission, the lawyer on a contingent fee." But some reformers argued differently.

In 1903, Ernest Poole argued that such work was not "a capital school for industry and enterprise," but that instead "The homeless, the most illiterate, the most dishonest, the most impure—these are the finished products of child street work." Lamenting street work's "unwholesome irregularity," he blasted the telegraph companies for their high turnover rates: "One of the large New York companies employs one thousand boys at a time, but employs six thousand during the year." Poole invoked the dead-end job argument, saying, "The best of them end as messengers on Wall Street. But even there it is rare to find one working up."[18]

Chapter 6 discussed how reformers involved with the National Child Labor Committee (NCLC) had fought the night telegraph messenger's exposure to "vice." But the NCLC also argued that messenger work was a waste of time for a child. General Secretary Owen Lovejoy rejected Western Union's claim that a messenger boy could learn telegraphy, quoting one messenger as saying "You can get to be a check boy or a file clerk after being in the service a year or two if you are wise and stick to it; but ordinarily nobody wants to stick to it." The problem, Lovejoy said, was that when not on call, messenger boys were forced to "loaf" within shouting distance of the manager. NCLC Ohio Valley secretary Edward Clopper agreed, asserting that "the messenger service is a blind alley; it leads nowhere." Chicago social workers Sophonisba Breckinridge and Edith Abbott explicitly linked together fears of delinquency and dead ends in their 1912 study of Chicago Juvenile Court cases, writing that messenger work, newspaper selling, and errand work were "'blind-alley' occupations which so often lead more or less directly to delinquent paths."[19]

But whether for reasons of danger or dead ends, reformers who focused only on the employment side of the problem were faced with few options. They could push for across-the-board restrictions on child-labor, but these were difficult to pass and impossible to enforce—the NCLC lobbied unsuccessfully for a child labor amendment through the 1920s. Reformers could hone in on specific excesses such as "night messenger work," but this targeting of exceptions gave implicit approval to the rule. And the more they actually succeeded in exposing the very worst child-labor situations, the harder it was for reformers to convince the public that there was still a child-labor problem. By 1923, the NCLC's director of research, Raymond G. Fuller, found himself redefining the problem with child labor not as brutal excess but as "interference with health, play, schooling," and other "normal" childhood activities. Instead of targeting ten-year-old factory workers, reformers targeted fourteen-year-old high school dropouts. An age-stratified argument against messenger work emerged: "For little boys, it is too hard, with too much bodily exposure and danger. For big boys, it isn't hard enough. It is a regular lazy-bones' job for a big, husky lad, with its hours of loafing and its irregularity of work."[20]

With this new targeting of older children, the focus of child-labor activism shifted away from workplace rules and instead to educational opportunities. Reformers advocated grade school and high school education in hopes of keeping children away from undesirable jobs. The New York Child Labor Committee pushed through a set of four state bills in 1903, the last of which kept children in school until age fourteen (instead of the previous age of twelve). These laws required work certificates to prove age and good health, but the rules were still easily evaded, and in any case nationwide in big cities only 47 percent of children who entered public schools remained after the legal age of fourteen. Instead of criminalizing truant children, reformers set about to change the schools themselves.[21]

Many reformers assumed that students who left school for work were different than students who stayed, in that the school-leavers were ill suited to a traditional classroom education; thus, they reasoned, the education system would have to adapt to these (allegedly) less intelligent or less motivated students. The solution offered in 1907 was a German-style "continuation school," defined as "any type of school which offers to people while they are at work opportunity for further education and training." This could mean evening schools, YMCA/YWCA schools, correspondence schools, or even schools run as employee welfare programs by private corporations, but all of the solutions involved changing the timing, the content, and the location of public education.[22]

At first, changing the timing of education meant schooling at night. Evening schools had been around since the late 1860s in New York City, and by the turn of the century some 180 cities had at least one free public evening school, with total U.S. enrollment at 292,000. The typical evening school ran two hours per evening for four evenings a week over twenty weeks of the year, giving a total of 160 hours of education per school year (regular schools had six times as many hours per year). Attendance on any given night ranged from only 20 to 60 percent of enrollment, and fully 85 percent of evening school pupils had not yet passed elementary school.[23]

Evening schools had originally been designed for working adults, but they were slowly transformed into institutions for working teens. In 1910, New York passed a state law forcing all boys under sixteen who hadn't graduated from elementary school to attend evening school for six hours per week, sixteen weeks per year. But out of a target population of 22,000 boys, only 7,000 registered, and only 3,000 ever attended. One high school principal said that the law failed for three reasons: parents and health officials objected to the late hours, children were too tired after work to learn, and employers were uninvolved. The law was repealed in 1913 and replaced with a new one: the same children had to attend school part time for four to eight hours per week, thirty-eight weeks per year, but attendance could now be during the day as well as at night.[24]

Little was done with this law by the New York City Board of Education until federal dollars became available through the 1917 Smith-Hughes Act. This measure mandated that at least one-third of all federal money going to states for teacher salaries in trade, home economics, and industrial subjects be spent on part-time continuation schools. The following year, the New York City Board of Education established its East Side Continuation School, the school that would later partner with Western Union. And in 1919, the state passed another law, raising the mandatory continuation school age to seventeen, since only about 12 percent of all boys in the New York City area aged sixteen to eighteen remained in school. Thus New York entered the 1920s with both a pool of money for continuation experiments and a large, legally defined population of continuation students.[25]

Besides changing the timing of education, the content of education in the continuation schools changed, becoming education about the world of work. The "vocational education" movement, dating from the work of Boston reformer Frank Parsons in 1907, offered a professional solution to the blind-alley problem: if students wanted to work at jobs and not at school, they should learn while working, in a job suited to their interests and abilities. With this goal in mind, New York City reformers began to explore the occupations open to children. A 1911 study, covering 132,000 working children aged twelve to fourteen, concluded that most youth jobs were blind alleys. A decade later, the U.S. Bureau of Education agreed, estimating that 87 percent of children who left school to work occupied dead-end jobs. Reformers feared the boy who "drifts from one job to another either because he does not like the work or because he has not had the proper training and not enough schooling . . . until the day when he realizes that he is a full grown man who has missed his calling." Franklin Keller, East Side Continuation School principal, pleaded, "Boys who are engaged in illicit trades and are on the highroad to crime and ruin, boys of brilliance and character who are being crushed under the weight of economic slavery, boys of power and endurance wasting their days and years in the blind-alley job, boys reared amid vileness and obscenity who will poison the society in which they live—these boys need the continuation school and its guiding, corrective influences."[26]

Thus, by 1922 the offical goals for continuation schooling in New York State were threefold: "(1) physical well-being, (2) economic independence, and (3) knowledge of the fundamental arts of reading, writing and arithmetic." The more instrumental, employable skills came first, with academic subjects limited to "setting up minimums." For example, in 1918 the Macy's department store continuation school in New York City tutored boys and girls aged fifteen to twenty in "arithmetic, spelling, reading, local geography, and hygiene," but only so students could "decide for themselves whether they are to become sales clerks or office workers."[27]

Through such private partnerships, the continuation schools ultimately changed not just the timing and content, but also the location of education. The University of the State of New York, empowered to oversee the 1919 continuation school law, drew up regulations allowing "private manufacturing or mercantile establishments and factories" to hold classes of their own for their young employees. Corporations had to provide a classroom and a workroom, plus teachers of comparable quality to those of the public schools, paid either by the corporation or by the local board of education. But teachers employed for twelve hours or fewer per week needed only to be "persons with a good general education who have some professional training or excellent technical training or extended practical experience"—in other words, any employee of the corporation.[28]

The university specifically detailed what these corporate continuation schools could teach. Of the four class hours per week, two could be used for "instruction in the practical work of the manufacturing or mercantile establishment or factory." Another half hour had to be used to teach American history, citizenship, industrial history, economics, and "laws relating to the industry taught." Finally, an hour and a half had to be spent on English, mathematics, hygiene, drawing, and science—but all "related to the occupation taught," except English. Thus practically the whole curriculum in these schools could be molded to corporate training goals.[29]

Private firms had already been experimenting with education by this time. In 1907, Wanamaker's in Philadelphia taught its younger cash boys on site, two mornings a week, in basic skills like math and writing. By 1913, the National Association of Corporation Schools, founded by New York Edison and National Cash Register, promoted this kind of education as independent of public school reform efforts. But at the same time, the National Association of Manufacturers began to argue for continuation schools in partnership with local school boards. By 1918, one researcher counted twenty-eight U.S. companies offering continuation schools, several specifically targeted to "boys sixteen to twenty" or "office boys." The New York City school board soon took corporate education even further, establishing a new Central Commercial Continuation School in 1925. The business community was closely involved with this school, serving on "a special advisory committee" to help with school management.[30]

The focus on vocational skills made it attractive to have businesses involved in the continuation schools. But allowing corporations to run their own schools on site was a more radical step. As early as 1912, NCLC authors had pleaded for using continuation schools to teach not only vocational skills, but also "such instruction as will make the young people more generally intelligent as citizens, better able to judge of public questions, with knowledge of and a taste for enjoyments of a superior character." In 1915, the left-leaning Commission on Industrial Relations argued that private company schools were too restrictive in the skills they taught, and advocated public-site continuation schools instead. And the U.S. Bureau of Education warned in 1921 that "the work done in schools connected with industries should be under public supervision, so that the technical instruction may be balanced by the training for broader and better citizenship."[31]

Despite such concerns, something had to be done to stem New York State dropout rates; by 1922, only 24 percent of sixteen-year-olds were in full-time school. New York City continuation schools blossomed in the early 1920s, quickly rising from an average daily attendance of about one thousand in 1920 to over three thousand in 1923, and private continuation schools were key to this growth. After all, the city public school system was in shambles. Even though the New York City Public Education Association had just spent $12 million to construct sixty-nine new school buildings, in 1905 two-thirds of New York City schools were so overcrowded that children attended only part-time, with schools running double half-day shifts and even seating children two per desk. The most crowded schools were in immigrant districts, whose older schools had originally been built for much smaller local populations. By 1914, New York City schools were still overcrowded: out of 800,000 students, almost 120,000 either attended double sessions or only attended part-time. This shortage of space meant that continuation students were given the worst accommodations available, "totally unsuitable buildings," according to one principal.[32]

The public schools also had a hard time managing truancy. No matter how easy reformers made it for working children to attend school on their off-hours, many more children registered than attended, desiring only an employment certificate. In 1915, out of 22,000 New York City children who should have been attending evening school, under 20 percent were registered, and only 9 percent attended. In 1923, some 4,436 students were registered in New York City continuation schools, but only 70 percent were attending, as opposed to 91 percent of day school registerees. And children commonly registered for continuation school without even having an official job; they spent four hours in school, and had the rest of the week to themselves (time that may well have been spent working for a wage, but not in an approved continuation-school job). Corporate continuation schools provided some insurance against this sort of truancy.[33]

Thus the Progressive campaigns against the excesses of capitalist child labor and dead-end jobs ended up right back where they started: at the doorstep of the corporation, under a new ideal of on-the-job vocational training through continuation schooling. What better test of such a turnaround in theory could there be than the single largest employer of child labor in the nation, Western Union?

Telegraph messengers were among the intended targets of the vocational education movement from the start. According to one Boston reformer in 1915, "almost the only way either a newsboy or messenger could get any vocational training was to stay in the trade long enough to get arrested and 'sent up' to a reform school." Confronted by such arguments, the telegraph companies fell back on their myths of internal and external advancement. In 1915, the Postal Telegraph journal ran a regular column entitled "Messengers Who Have Advanced," each spotlighting a current PT manager who had started out as a messenger. Terre Haute, Indiana manager Alwyn J. Doyle, readers were told, started as a messenger at age sixteen in 1902, and, "Like nearly all the bright boys who enter telegraph offices as messengers, Alwyn at once began fooling with the keys, and in the course of time became an expert operator." A Los Angeles cashier started as a messenger in 1893, "so small that when he went into a customer's office the customer was likely to look around to see where the boy's mother was." And in a sobering example of persistence, Philadelphia manager James Wilson started at age twelve as a messenger in 1877, and became a manager thirteen years later after working as office boy, clerk, cashier, and operator.[34]

It wasn't hard to find individual cases of success among the messengers; but few of these celebratory articles discussed the sheer number of boys an office would churn through before it found a hero worth writing about. One editorial counted that in New York City from 1912 to 1915, 146 Postal Telegraph messengers had been promoted, or about 50 a year: 107 to clerk, 25 to operator, 9 to check-boy, 2 to lineman, and 2 to manager. Postal crowed that it was "doing a work for boys that is greater than all the work that could be performed among them by any Welfare Society." But in 1908, PT had employed 612 messengers in New York City, and that number had most likely risen by 1915. Even if it hadn't, the promotion of 50 out of 612 messengers—only 12 percent—was hardly something to brag about, especially since ten times the number promoted had likely left the service that same year.[35]

The biggest audience for such tales of advancement was the pool of prospective messengers themselves—with such high turnover rates, telegraph companies needed to continually attract new boys. In 1918, a Seattle Western Union office advertised, "Western Union service offers a fine business training for boys. A branch of the Public Library is installed in the office for the boys' use. Telegraphy taught free of charge during employees' hours." The "free" training suggested that given such opportunity, failure was surely the fault of the boy. At PT, even extra work was to be seen as opportunity, not oppression: "When [the manager] requests you to do odd jobs around the office he has no selfish motive; he is looking for a promising telegraph man. Therefore, you can accept it as a compliment."[36]

But the reality of messenger employment was different, and the telegraph companies knew it. Boys at the bottom would rarely rise to "telegraph man"—but the telegraph companies, particularly WU, still needed to attract those boys. Thus the push for vocational guidance through continuation schools offered WU a unique opportunity: the company could recast its messenger department as a social service and set up a school for messengers in its own building, both answering the child-labor critics and receiving the full support of the state.

The Western Union Continuation School was approved by the New York City Board of Education in September 1923. Classes were to be held at the company's 395 Broadway building as an annex of the West Side Continuation School (oversight was eventually shifted, first to the Central Commercial Continuation School, and later to the East Side Continuation School). The Board of Education explained their reasons for approving this new school in their minutes, using two arguments. First of all, it was more efficient to consolidate Western Union's "five to eight hundred boys" in terms of attendance tracking: "At present these boys attend seven different continuation schools, and it is impossible for the company to fully cooperate with the Board of Education in looking after the attendance." Second, the new school would save the school district money, as "The company offers well-equipped premises without cost and is also willing to pay the salaries of fifty per cent. of the teachers." All in all, the board concluded that "The offer of the Western Union Company is in line with progressive tendencies."[37]

The school began immediately. Five elementary classes per week were held in the mornings, and a high school class was held in the evenings. A few years later, in June 1926, the school graduated its first class. There were so many students that the graduation was held in the city's town hall, but "without cost to the Board of Education." That year, WU would proudly proclaim its new mission, calling messenger work "The Bridge Between School and Business," as well as "the prep school to the University of Business."[38]

That "business" did not mean telegraphy. The messenger continuation school was not an operator's school; WU had separate schools "for the purpose of training those who are to become either automatic, Morse, or telephone operators for the company," both men and women. Nor was the messenger school a training ground for future WU managers; in 1925 and 1926, WU spent over $13,000 to set up a special manager training school in Tyler, Texas. The messenger continuation school did not fit the myth of internal advancement within telegraphy. Instead, it was to train students in other vocations.[39]

If not to grow its own employees, why would WU bother with such a school? First, it provided a legal mechanism whereby WU could continue to employ boys aged fourteen to eighteen during the working day, while maintaining control over their time and activities. Second, it served an advertising purpose: the Board of Education, eager to have WU supply space for even a fraction of its boys, spoke publicly of WU's good deed. Third, WU hoped that by engaging the boys in long-term educational activities they would stem the tide of messenger turnover. And finally, WU hoped they might even build better messengers in the bargain, producing "more efficient work" and a consequent "reduction in the cost of delivering messages."[40]

Because of the temporality of messenger work, an on-site school also addressed WU's load-leveling needs. The Board of Education allowed WU to bend the continuation school rules and hold one of their five classes per week on Saturday mornings from 8:00 A.M. to noon instead of on Monday mornings, because Saturday was a much slower day for WU than Monday. WU general manager J. F. Nathan himself pressed for this change, arguing "we are straining every energy to bring the attendance at our continuation school up to a high mark," but cautioning that "we are doing so at no small sacrifice of service to our patrons, the public." He characterized the continuation law as a "hardship": "our obligations under the law . . . perhaps fall heavier upon us than upon other employers because of the large number of boys in our employ."[41]

Both WU and the messenger school prospered through the 1920s. In 1928, WU even spent $5,000 for "additional equipment and for rearranging space," and began referring internally to the school as both "continuation school and employment bureau" (though it was not legally registered as such). Western Union had the largest single continuation school in New York City, with one thousand boys registered. Elementary students were divided between industrial courses (radio, electric wiring, electrical installation, woodworking, home mechanics, and industrial practice) and commercial courses (industrial and commercial history, economics, hygiene, civics, typewriting, and mimeographing). As for the high school enrollment, classes included bookkeeping, typewriting, shorthand, history, geography, mathematics, and English. There were also music classes and a Western Union Messengers' Band of sixty musicians, all under age seventeen.[42]

In 1928, the school moved into the new WU high-rise office building at 60 Hudson Street. Originally, only operating activities were to move to the new twenty-four-story skyscraper, with four thousand WU employees to occupy six floors. But space was made for the messengers as well:

In accordance with the Company's policy of caring for the physical and mental welfare of its messengers, one entire floor of the new building will be devoted to the messenger service. There will be classrooms for continuation and high school work, study rooms, library, locker rooms and uniform depot. There will also be completely equipped workshops in which the boys will be taught electrical wiring, printing, plumbing and heating, automobile repairing and sheet metal work, thus giving any one of the 1,500 messenger boys in New York City ample opportunity to decide if he prefers a trade to any of the several kinds of office work with which he comes in contact as he delivers telegrams.

There was also an auditorium that seated one thousand, to be used for messenger safety meetings, school commencement, and other public events (quite a turnaround from the days of hiding messengers in the basement).[43]

The claims WU made about its "apprenticeship" now grew more grandiose. By 1928, general manager Nathan was speaking of the molding of future businessmen as the real goal of the entire messenger service, with Western Union acting as "a kind of clearing-house for high-type boys who are attracted by the opportunity to earn money and at the same time get the benefit of a vast variety of business and industrial contacts which will enable them to look the field over well before deciding what branch of business they wish to enter."[44]

Such rhetoric succeeded in stemming the tide of criticism from major child-welfare groups in the mid-1920s. The NCLC, which had so vehemently attacked WU for allowing its messengers to be exposed to "vice" in the 1910s, ignored WU messengers in the 1920s, recategorizing the dreaded category of "street traders" as simply newsboys, bootblacks, and vendors. With the Boy Scouts of America, Western Union used more direct public relations tactics. Between 1923 and 1928, WU regularly purchased advertising in *Boy's Life*, and WU president Newcomb Carlton himself served on the executive board of the organization. Western Union also earned (and publicized) the support of the YMCA. In a 1926 messenger recruiting pamphlet, WU reprinted a letter from the head of the Wall Street YMCA that parroted the messenger advancement myth, calling the company "a prep school of business where boys come in for a two or three year course and then go on, filling higher positions within their own organization, or stepping out into another business." And education reformers warmed up to WU as well during this period, with one author calling messenger work "much more than a blind alley job."[45]

Through its messenger school, Western Union claimed to fulfill the promise of the public education system itself—only better. Their rhetoric reached epic proportions in 1932, with WU invoking both mechanical and military metaphors in likening itself to "a gigantic grist mill into which is poured each year an army of boys, averaging fifteen years of age with no business experience and usually no fixed objective in life." The company claimed that this "raw material" would "emerge from messenger service with a more comprehensive knowledge of modern business, a fundamental groundwork of human character, and a more thorough understanding of life's possibilities."[46]

But how well did the reality of continuation schools match the rhetoric?

The state worried that it cost more to educate children in continuation schools than in conventional schools, even with corporate partnership. New York City superintendent of schools William Ettinger admitted that the cost per pupil of continuation schools in 1921 was 28¢, while for day school students it was only 9¢. But he justified the higher cost by imagining the economic benefit to the city of having so many children at work—over three thousand in 1924, each with an alleged average wage of $649 per year. Thus the fear that special education for poor children would be more costly to all was turned on its head: the labor of poor children was cast as a benefit to the entire economy.[47]

A year later, a committee surveying New York City's vocational and continuation schools concluded, "There is need for a thorough going [sic] revision of the whole situation." The biggest criticism was that continuation schools didn't result in higher wages for their students as promised. A 1921 survey of the weekly wages of two hundred continuation boys aged fifteen to seventeen revealed little change as the boys moved through the new educational program. A 1923 NCLC report on 8,517 New Jersey continuation school students found that the median weekly wage for such boys was $9.56, arguing, "They leave school for work that pays the most and not for that which promises the best future." And a 1925 report claimed that the eventual weekly earnings of children who left school at age fourteen, even with continuation classes, were much less than earnings of those who left school at age eighteen—a difference, once the child reached his mid-twenties, between making around $10 per week as a grammar school graduate, and $20 to $30 per week as a high school graduate.[48]

Moreover, continuation schools were always tarred with the stereotypes of their students. The original 1907 report on continuation education characterized pupils who remained in the regular day school system as "the brighter element" compared to those who dropped out. This was certainly not the only way that contemporary investigators conceptualized dropouts. When Chicago factory inspector Helen Todd surveyed five hundred working children to ask whether they'd rather go to school or work, over 80 percent said they preferred factory work, not because of their own intellectual limitations, but because they characterized school as a crueler and more stressful place. Yet the image of the dropout as mentally weak persisted, buttressed later by dubious intelligence tests. Many in the vocational movement thought that certain students were inherently "fitted" to certain occupations, and that schools had to be reformulated to provide a range of skills and assistance to students based on a single estimate of their potential.[49]

Messengers were tainted with such prejudices too, as they made up a large proportion of New York City continuation students—nearly half of the 308 boys who got their first employment certificates from the East Side Continuation School in 1924 were classified as "messengers" (though these probably weren't all telegraph messengers). Despite the advertisements of the telegraph companies, messengers were often characterized as slow, stupid, and even criminal. The NCLC's Clopper cited a report that showed that 50 percent of "delinquent" children brought before the juvenile courts were street workers, and that "those who are attending school and doing street work at the same time constitute quite a large percentage of the pupils who are retarded."[50]

Even school officials shared this view of messengers. East Side Continuation School principal Keller wrote that continuation students who worked at "errand" jobs were "doing work requiring no more intelligence or skill than that possessed by a high-grade moron." When discussing "the subnormal juvenile worker," Keller pointed to the messengers specifically, recounting a long list of jobs culled from "an examination of the records of employment bureaus" showing where "subnormal" men and boys were employed. First on the list? "Telegraph messenger." Other such jobs, for comparison, were delivery boy, errand boy, peddler, newsboy, operator on machine, helper, porter, bootblack, elevator operator, "pin-boy in bowling alley," "hat-boy in club-house," and "nailing flags to sticks."[51]

Clearly, as illustrated earlier, messenger work required more skill than did nailing flags to sticks. Messengers had to be able to read and write names and addresses; to use maps and understand geographic directions; to calculate time, distance, and speed; to add and subtract tariffs and fees; and to understand, remember, and explain telegraph regulations. Messengers used problem-solving skills in order to make deliveries with incomplete or incorrect addresses, to track down customers between their homes and offices, and to sell quotas of holiday telegrams. And messengers who were successful in for-hire assignments would have to know how to operate office equipment, mind children, and perform a wide range of skilled tasks. But as continuation students, messengers were thought to be unsuitable for other employment, in direct contradiction to WU's own proclamations about the bright futures ahead of its boys.

Nevertheless, the WU school survived through all of these criticisms, even as other continuation schools were folded back into the normal daytime curriculum. Yet WU had its own questions about the effectiveness of its school in stemming turnover. After all, with the relative economic prosperity of the 1920s and the success of corporate paternalism benefits, manufacturing employee turnover rates had fallen through the 1920s to 30 percent per year by 1929—down from nearly 100 percent in the early 1910s. But turnover at WU was just as bad as before. Even though WU messenger manager Fowler claimed in 1926 that there was "nothing to worry about in messenger turnover," and that it was "a healthy condition," the continuation school threatened to become a liability, as WU spent money to enroll, process, and educate boys who weren't even going to stay around to work for them.[52]

Figure 8.1. *Western Union messenger employment/education, New York City, 1926–1932.*

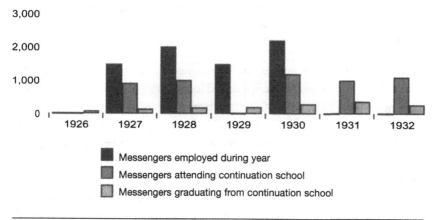

Messengers employed during year
Messengers attending continuation school
Messengers graduating from continuation school

Sources: *T Age* (1926–32); Dreese (1929); *Elementary School Journal* (June 1928); *D&D* (1932–37); *NYT* (1928–42).
Note: Blank values indicate missing data.

Consider the numbers: WU advertised that it "graduated" one hundred to three hundred messenger boys each year, but most of these boys received only an elementary school diploma, not a high school certificate. And in a city that demanded fifteen hundred to two thousand messenger boys at any given time, only about half of those boys attended the continuation school. Then there was the turnover effect. In 1928, for example, to fill those two thousand messenger slots, some eight thousand boys had to be hired that year. These turnover numbers were reflected in the continuation school as well: even though one thousand students attended the school at any one time, fully five thousand might be enrolled in the course of the year[53] (see figure 8.1).

Around 1926, WU commissioned Columbia University graduate student Ira Dreese to conduct a "scientific management" study to help the company decide how it could best achieve the somewhat contradictory goals of both stemming messenger turnover and avoiding long-term messengers. Dreese's review of employment records confirmed that from 1919 to 1928, monthly turnover in New York City averaged 30 percent—meaning four times the number of active messengers were "released" every year. Breaking this turnover down, Dreese found that much of it was rapid, with a core group remaining on the job. Henry Miller's observations a few years earlier corroborated this assessment:

Perhaps twenty per cent of the force was steady; the rest was driftwood. The steady ones drove the new ones away. The steady ones earned forty to fifty dollars a week, sometimes sixty or seventy-five, sometimes as much as a hundred dollars a week, which is to say that they earned far more than the clerks and often more than their own managers. As for the new ones, they found it difficult to earn ten dollars a week. Some of them worked an hour and quit, often throwing a batch of telegrams in the garbage can or down the sewer.

Dreese found that night messengers (often men over age twenty-one) were the most stable, with 44 percent having served longer than one year; the few female messengers were next, with 36 percent having served more than a year; and finally came full-time day messengers, 23 percent of which had served over a year. Ironically, the rest of his study focused on these daytime boy messengers, ignoring the obvious conclusion that if WU really wanted to stem turnover, it could simply hire more adults and women at higher wages.[54]

Why did the boys leave? Western Union vice president Fowler had claimed that each year WU promoted nine hundred messengers, but through a study of two separate months of messenger "releases," Dreese found that only 1 percent or so were promoted within WU. Around 19 percent were fired, but only 4 percent for on-the-job reasons—the other 15 percent were "fired" for not showing up. Some 32 percent left because they found a better job. The rest, 48 percent, simply disliked messenger work, mostly giving reasons of "low pay" and "too much walking." So nearly half of the turnover (and possibly more if the 15 percent no-shows are included) wasn't related at all to "advancement," but was due to poor working conditions and wages. Such boys left, according to Dreese, because "youth with its restlessness and freedom from economic responsibility is a continuous period of vocational try-outs."[55]

Dreese's point of view was not unusual. The principal of the East Side Continuation School himself saw child turnover as the child's psychological problem, not a problem of the employer, arguing that "regardless of continuation school, many boys and girls flit from job to job. In technical parlance they are job 'hoboes.'" This sort of tautological reasoning—children "flit from job to job" because of "a lack of stability"—served only to legitimize the conditions that led to high turnover, and to ignore the agency exerted by child workers in seeking better employment situations.[56]

But interestingly, WU didn't want to entirely end turnover; it only wanted to have more control over turnover. Western Union wanted messengers to stay more than three months, but less than one year. These endpoints were not arbitrary: with messengers who left before three months (a full 75 percent of all messengers they hired), WU lost money in hiring, processing, uniforming, and training them; with messengers who stayed after one year, WU lost money paying them raises and allotting them vacation time. And given the propensity of messengers to strike (discussed in chapter 9), WU also probably feared that the longer a messenger stayed on the job, the more militant he was likely to become, and the more knowledge he would have about ways to thwart the WU system.

Dreese's report put a numerical value on WU's turnover losses. He estimated that hiring a messenger, including advertising, interviewing, and processing, cost $2. Fitting him for a uniform cost 50¢. The cost of a few hours of "classroom" training (not continuation school, just the rules of the job) was another 50¢. Finally, the first probationary week of work, paid at a generous flat wage instead of a low piece wage, cost an estimated $3.50 per messenger. Thus, the total hiring cost for each messenger was $6.50. Using these figures, Dreese determined that the 3,880 messengers hired in 1928 who were "released" before three months cost Western Union $20,000.[57]

Dreese concluded that WU hiring policy, not messenger schooling, was the key to stemming turnover. First, he said, WU should abandon its existing application and interview process, since most of those who were rejected for messenger work were rejected for nonscientific reasons of "poor messenger type" (30 percent), "previous employment record," (24 percent) or physical appearance (21 percent). Dreese laboriously correlated messenger "success"—staying on the job more than three but less than twelve months—with everything from nationality to IQ, and advised WU to seek out "younger, non-Jewish boys with an education of ninth grade or less," from social agencies in particular, to get the optimum length and quality of service. In addition, he recommended bonuses to entice boys to stay three months, and thought "an educational campaign advising boys how to take proper care of their feet might aid in lessening resignations due to 'too much walking.'" Thus Dreese concluded that the continuation schools had no effect on WU messenger turnover, advising education only in how to walk, not in how to find a career.[58]

Apparently WU did try to change its employment practices as a result of Dreese's study. According to WU personnel manager E. A. Nicol, "In many cities trained personnel men are being employed to increase the standards of selection, supervision, and follow-up work with messengers." Nicol also stressed changes in the physical environment: "Emphasis has been placed upon the methods of interviewing and selecting boys, the quarters into which they are ushered in their initial contact with the company, and the reception of each boy, both at the employment center and in the branch office." (He didn't reveal whether or not WU was moving to Dreese's non-Jewish, grammar-school force.)[59]

Nevertheless, turnover continued apace. In 1930, WU bragged that out of 15,000 total messengers across the United States, 286 were promoted within WU, 465 were "placed" by WU as a result of calls from WU customers requesting "junior clerical help," and 2,325 others supposedly found new jobs due to their messenger contacts. But this total of 3,076 is less than one-tenth of the estimated 50,000 boys who cycled through WU that year to fill those 15,000 slots. The rest were "either unfit or disinterested," according to Nicol. Oddly, those "unfit or disinterested" boys were still part of WU's advertising statistic that same year, when their article "Messengers Today—Business Men of Tomorrow" proclaimed, "approximately 50,000 boys each year are graduated from the great 'school' of business contacts, personal service and company training known as the Western Union messenger service."[60]

The continuation school experiment lasted only as long as relative prosperity allowed. Across the United States, the number of continuation schools rose to a peak of around 337 in the mid-1920s, with some 350,000 students officially enrolled. But both school numbers and enrollments started to decline with the Great Depression, as available jobs were increasingly taken by adults and fewer employment slots for children existed in continuation schools. New legislation helped keep children in day school to age sixteen, and a greater demand for adult-level evening courses taxed the school systems even further. Most continuation schools were folded back into commercial or vocational high schools by the late 1930s.[61]

But WU's continuation school lasted long after others had faded, encouraging messengers and customers alike to put a little more faith in the framed portrait of Andrew Carnegie that hung in every WU branch office over the words "To our messenger boys of which I was one." In 1932, WU's Nicol told the story of "Alfred," a "small and somewhat unkempt Italian boy" who came to WU after graduating from grade school. Western Union set him up with a job, gave him "an understanding of the values of personal appearance," put him in continuation school, and offered him a chance to join the Boy Scouts. Two years later, Alfred was transformed, having been offered three different jobs by companies he delivered messages to. "I doubt if any guidance worker could have accomplished with Alfred what he did for himself as the result of his work experience," said Nicol. Individual stories like this—both real and fictionalized—were still WU's best defense against the statistical reality of their huge messenger service.[62]

As late as 1952, the Western Union messenger school was still in operation. But by this time, WU wasn't making such grand claims about its own "business education," instead acknowledging that a true public education should come first: "Before a messenger of school age is given employment, his scholastic record is checked with school authorities and their recommendations obtained. Subsequent contacts at frequent intervals are made as a precautionary measure against the possibility of interference with the messenger's studies." Nevertheless, the 1950 census showed that out of a total of 8,388 telegraph messengers counted nationwide, only 1,806 had completed high school—or about 22 percent of all messengers. Western Union rarely practiced diligent monitoring of messenger grades. But even the rhetoric was a shift from the 1920s, when WU claimed to educate boys better than the public school system itself.[63]

Western Union argued that its continuation school and its messenger service together made boys into men, a career and education myth that worked right through the worst years of the Depression. By 1938, the American Communications Association, the first union to seriously court the messenger boys, would argue that the labors of boys did *not* lead to the labors of men, as it tried to convince the U.S. Children's Bureau that the labor of "23,000 boys in America who today are earning from $6.00 to $8.00 a week for a workweek from 48 to 60 hours" should be considered "hazardous" and outlawed. The next chapter will explore how this sort of union activity, and not the efforts of the child-labor and education reform groups, finally made the messenger boy into the messenger man.[64]

CHAPTER NINE

FROM UNION JOKE TO UNION MAN

There has never been, we believe, a single instance in history of a strike caused or threatened by the employment in commerce of messengers under sixteen, and there is no basis for any fear that any interruption to commerce ever will be so caused or threatened.

—Western Union assistant vice president T. B. Gittings before
the Senate Committee on Education and Labor, 1945[1]

All that the boys want is fair play and a fair chance to live, and, in return, they give honesty, industry and sometimes their health.

—Letter from a striking messenger boy,
Operator and Electrical World, 1883[2]

During the early years of telegraph expansion under Western Union, there were four operator-only unionization attempts, all of which ended in failure. When confronted with a new union, WU was always able to offer comparable union benefits itself, such as journals and insurance programs; when confronted with a strike threat, WU was always able to use its contractual links to the railroad and press companies to call in replacement labor. For their own part, the commercial telegraph unions had trouble reaching out to operators across other business units such as the railroads and the press. And early union operators also ignored their young coworkers, the telegraph messengers—even though those messengers had a long history of labor militancy of their own. As this chapter will show, only after the adult, skilled telegraph workers began to ally with their youthful, unskilled counterparts in the 1930s would the unionization movement in telegraphy have any real success.

By the early 1860s, there were numerous reasons for telegraph operators to want a union. During the Civil War, operators faced long hours, poor pay, and "small, dark, filthy, and poorly equipped" offices. Even though demand for telegraphers was high, control of the craft was difficult, due to the "plug schools." But instead of a militant labor union, New York City operators set up the National Telegraphic Union (NTU) in 1863 as a "mutual benefit society" providing sickness and funeral payments and its own journal, the *Telegrapher*. The NTU excluded all other telegraph workers, including messengers, and did not demand better wages; it was neither a secret organization, nor was it opposed by the telegraph companies. But by 1866, after the WU merger with American Telegraph and United States Telegraph, NTU members started to complain of lost jobs, lowered salaries, and canceled vacations, and began to push for a more oppositional union.[3]

Western Union president William Orton responded to the union threat by matching labor's offerings with similar company benefits (a tactic WU would use half a century later in establishing a company union). To respond to the *Telegrapher*, Orton started the *Journal of the Telegraph*. Editor J. D. Reid noted, "With [my] easy chair in the Executive rooms of the Western Union Telegraph Company, it will not be expected that [I] will say anything very disrespectful of its management." Similarly, in response to the NTU's death benefit, WU created a Telegraphers' Mutual Life Insurance Association, with Reid as treasurer.[4]

As operator conditions worsened in the late 1860s, many NTU members demanded a new union willing to strike. New York delegates to the second-to-last NTU convention in 1868 formed a new organization, the Telegraphers Protective League (TPL). Still exclusively for operators, this secret group pressed for job concessions. The TPL was tested in 1870 when San Francisco union operators went on a wildcat strike after threatened wage reductions and the firing of a union officer. The strike spread to thirty-five cities, but failed within a week, for several reasons: it occurred during the slow season (especially slow for news, since Congress was not in session); the TPL was too weak in numbers and in finances; and WU could easily bring in "scab" operators from the railroads (who were in their own slow season). The TPL quickly disappeared.[5]

Reid later claimed that "Up to the dawn of 1870, labor in all departments had been performed not only without murmur, but with devotion and often with heroism." Murmurs or not, the 1870 strike helped to make labor-management relations more oppositional in the years to come, as operators responded to WU rules through "informal resistance and struggle" such as slowdowns, sabotage, and high job turnover.[6]

The next union attempt had a wider base. In 1869, a group of Philadelphia garment cutters founded the Knights of Labor (KofL) as a broad but secret union of all workers (though most members were skilled craft workers). A decade later, in 1878, the KofL counted some 9,000 members, both skilled and unskilled. The KofL proposed public lands for settlers, an eight-hour day, an end to child labor, and equal pay for the sexes, in an attempt to end "wage slavery" by encouraging cooperatives, not by striking against capital. Still, some 30 percent of their "resistance fund" was earmarked for strikes. Over the next few years, the KofL continued to be erratically popular, as membership grew to 28,000 but then fell to around 19,500.[7]

The KofL base, coupled with Jay Gould's sudden takeover of WU in 1881, motivated the next attempt at a telegraph operators' union. In 1882, two nascent unions joined as the Brotherhood of Telegraphers (BoT) under the KofL, whose membership by 1882 had reached a new high of 42,000. Organizing both railroad and commercial operators, by 1883 the BoT claimed 8,198 members out of a total of 22,200 operators in the United States and Canada. But rather than just an operators' union, the BoT aspired to be "an industrial union comprising all who created the telegraph companies' wealth, whether smartly dressed operators who manipulated delicate and temperamental instruments or rough-hewn linemen shod with much-encrusted climbing boots." Yet one class of wealth-creating employees was overlooked, despite having walked out over wages in New York City and Boston four times that year: the messengers.[8]

The BoT's first major action against the telegraph companies came quickly. In July 1883, operators and linemen across the nation went on strike against WU, demanding eighty-hour days, seven-hour nights, an end to Sunday work, equal pay for equal work (even across sex), better conditions, and a 15 percent pay raise. But WU general manager Thomas Eckert argued that the union did not represent the workers. The strikers found favorable publicity, and even garnered sympathy from New York City messenger boys, who reportedly "were not organized and didn't have any particular grievance to redress" but "felt the influence of secession and longed to join the larger army of strikers." Yet, as in 1870, both the press telegraphers and the railway telegraphers failed to join the effort, and the "Great Strike of 1883" was lost after a month. The BoT made one last attempt in 1886 to woo the railway operators, but those workers formed their own union the same year, the Order of Railway Telegraphers. By 1887, WU felt it had put the "labor question" to rest—as Reid put it, "We had better be content with the work given us which we are able to do, and perform it faithfully, than vainly wish for something beyond our reach which we would not be able, perhaps, properly to perform."[9]

All this time, despite the lack of attention from the operators and their unions, the messengers, newly reconstructed by ADT and employed in greater numbers after the 1874 ADT-WU agreement, began to grow their own form of labor militancy. Messenger strikes increased in frequency, size, geographic extent, effectiveness, and union involvement, all the way to the turn of the century.

Only a few months after ADT and WU reached their initial joint agreement in 1874, ADT boys in New York City struck over an increase in daily hours from eight to ten with no raise in their $4 per week pay. The strike resulted in four boys being fired and sixteen being fined for "leaving their posts." Similar small, unsuccessful strikes occurred in other places as well, such as in Boston in May 1875, when WU boys struck over uniform costs and rules. Such strikes failed, according to the *Telegrapher*, because of "the superabundance of labor, especially of juvenile labor, in our large cities."[10]

Yet the messengers had the power to paralyze the entire telegraph system, if only they would all strike together. The New York City ADT boys eventually tried this, staging their first mass strike in August 1880. Originally, around one hundred ADT messengers walked out for a wage increase. But the strike grew until around half of ADT's entire seven-hundred-strong New York City messenger force was out, affecting nearly all city offices. However, by bringing in strikebreakers from other companies and arresting any who interfered with the replacements, the company broke the strike within a week. As the *Telegrapher* had predicted, success could only come if the boys could keep the telegraph companies from quickly and easily hiring replacements.[11]

Perhaps because of the rapid turnover in the messenger force, failures did not dissuade messengers from striking year after year. Messenger strikes were part of the "Great Upheaval" of the 1880s, breaking out at district telegraph offices in various cities through 1881, including ADT Brooklyn in June and ADT Chicago in September. All the strikes failed, but they did earn some notice. The *Operator* commented, "the tender years of the lads may serve as an excuse for their conduct." The *New York Times* also dismissed the strikes as children's folly, writing that the striking boys were "very noisy" and that "very few, it is said, went into the strike with any heart for it."[12]

But no matter what the media declared, messenger strikes continued, and even began to succeed. A May 1882 WU messenger strike in Boston over a proposed pay decrease from 2.5¢ to 1.25¢ per message under a new WU-ADT contract was victorious. Tactics of the striking "trotters" included posting signs all over town warning other boys to stay away from WU and harassing anyone hired to break the strike. Western Union was forced to use linemen, clerks, and other employees to deliver messages. Some local businessmen sided with the messengers, writing letters to the telegraph managers. After three days, the company agreed to keep the current pay rate. This success hinged on the very resources that made messengers valuable in the first place: urban mobility, the sympathy of the local business contacts they served every day, and youthful energy. But victory was still rare in the larger messenger market of New York City. Even when 660 ADT New York City messengers went on strike a month later over the threatened loss of a 57¢ Sunday holiday, forcing the company to back down on the rule, the twenty-five boys who organized the strike were fired.[13]

The ease with which striking messengers were fired showed that while the boys might have exerted their right to protest, they still hadn't earned the right to organize. In early 1887, the still popular KofL became involved in a New York City messenger strike, the first union to do so. This was also the first time that messengers from both ADT and WU all joined in a single, citywide strike. The strike was unsuccessful, even with the KofL's assistance, and the alleged messenger leader was arrested but acquitted. The judge commented, "[The messengers] have as much right to meet together as Jay Gould and his friends."[14]

This strike revealed the difficulty that children faced when using adult tactics to demand adult wages. When ADT messenger manager Watson Sanford found out that 150 of his employees had been meeting with the KofL to list grievances, he sent a letter to all the *parents* of the messengers, warning them of the dangers of unionization:

DEAR PARENTS AND GUARDIANS:

We send you a notice that we are well informed from reliable sources that your son is in danger of falling into the hands of agents who are trying to bind together the honest working boys of this city and lead them to commit acts against the laws of peace, good order, and honest industry. We wish to warn him through you to be on his guard, to keep away from such influences, and avoid serious results. We beg to ask you to be on the watch and advise him to keep out of bad company and continue in the esteem of his employer as a good, law-abiding servant of the company.

Sanford claimed that more than seventy parents called to thank him for this letter, showing that the failure of this strike could not be attributed to capital alone.[15]

That the 1887 strike resulted in union involvement, cautionary letters to parents, and a judge's reprimand shows how seriously it was taken. But later messenger strikes would be seen as less and less important. By 1899, *Telegraph Age* was painting messenger strikes as the absurd actions of lazy boys who "refuse to allow, if strikes can prevent it, any arrogant and money-bloated corporation to compel them to hurry." Such strikes had become routine by 1910, when the *Times* wrote of "Wall Street's annual strike of messenger boys" and quoted an ADT manager as saying, "It is a sign of Spring, this strike talk among the messenger boys. It comes every year and nobody takes it seriously."[16]

Figure 9.1. *The press lampoons striking messengers, 1910*

THE HORRORS
OF THE STRIKE

Source: *NYT* (November 2, 1910).

Now even large messenger strikes were lampooned. In November 1910, after six hundred messengers attended a meeting with union representatives from the Central Federated Union, five hundred ADT boys from forty offices around New York City went on strike. They were soon joined by PT boys as well. But within a week and a half, the strike had been broken, since ADT and PT were willing to pay bonuses to the abundant strikebreakers. Still, the strike gained national press in the social work journal *Survey* and the attention of a major labor group. Messenger boys in Philadelphia, Pittsburgh, and Los Angeles went on strike as well, apparently inspired by the New York City boys. But the *Times* poked fun at the strikers in cartoons (portraying a messenger being spanked by his mother) and caricatured their public speeches as uneducated[17] (see figure 9.1).

In the same way, messenger strike actions, although increasing in frequency and effectiveness, were still of little concern to WU and PT, for two reasons. First, messenger labor actions were constrained in space. Unlike operators' unions (which could be organized through national journal communications) or operators' strikes (which could be signaled nationwide over the telegraph), messenger actions were inevitably confined to a single city—and often to a single neighborhood or office. Second, messenger labor actions were constrained in time. Unlike operators, who could plan on a relatively stable career with time to build a union over several years, messengers moved through the telegraph company in a revolving door, rarely remaining for over a year. Neither long-term gains nor a shared memory of struggle were possible within the messenger population itself.

Meanwhile, by the turn of the century, enough new operators felt the need for a union to make a fourth attempt at organization. In 1903, the Commercial Telegrapher's Union (CTU) was formed, comprising sixty locals and 8,010 members, with separate divisions for press, brokerage, and commercial operators. The new union affiliated with the American Federation of Labor (AFL), a group of trade unions counting nearly 150,000 members that had been formed in 1886, headed by Samuel Gompers until his death in 1924. Unlike the KofL, the AFL would not interfere in trade union activities, only guide and arbitrate among them.[18]

Messengers were again excluded from the CTU; however, the union mouthpiece, the *Commercial Telegrapher's Journal* (*CTJ*), talked of messengers regularly. By 1903, messengers were now striking not just in big cities like New York, but in smaller cities as well: a May WU messenger strike in Butte, Montana, involved one office; an October WU messenger strike in Memphis, Tennessee involved twenty-one out of twenty-two boys; a November WU messenger strike in Cambridge, Massachusetts resulted in an increase of $4 per month; and a December ADT messenger strike in Knoxville, Tennessee, was counted as a victory since the company agreed to consider a wage "adjustment." After fifty-three Portland, Oregon, messenger boys set up a "Messenger Boy's Protective Union" affiliated with the AFL, in order to "keep the lads off the streets and amused during leisure hours" with a club room, gymnasium, and reading room, the CTU took notice of the messenger boys as the unionized operators of tomorrow: "Some of these little fellows are beginning to realize that the only way to redress their grievances lies in organization. They are to be congratulated." At the same time, the CTU lamented the fact that some of the "men with the intelligence of the average telegrapher of today" did not themselves organize, saying "you are not doing even as much as the poorly paid messengers, who have had the manhood to organize for their own protection."[19]

When union leaders described messengers as future operators, they fueled the same myth of internal messenger advancement that the telegraph managers did (discussed in chapter 8). But telegraph operators were wary of messenger advancement rather than celebratory. By the early 1900s, operator salaries had dwindled, operators were forced to supply their own "mills" (typewriters), greater numbers of women operators were being hired at lower wages, and more and more "ham" operators were graduating from the plug schools. CTU operators feared that messenger boys trained by the telegraph companies would be offered existing operator jobs at lower salaries, dropping operator wages from

Figure 9.2. *CTU cartoon of a nonunion messenger, 1905 (detail)*

Source: *CTJ* (October 1905), 18.

$60 to $70 per month to $35 to $40 per month. They advised: "Discourage these would-be 'knights of the key' and tell them that the 'telegraph profession' is a thing of the past; that the supply is already greater than the demand."[20]

It wasn't just a lowered wage floor that these union operators feared—it was the idea that nonunion messengers would be promoted to take away union operator jobs, thwarting the long effort of building a union. One CTU cartoon showed a young, smoking messenger boy asking the "corporation-owned chief operator" for a job, with the chief replying, "Yes, get a mill and come around next week. I'll put you on extra [contingent labor]. Stay out of the union, read the [*Telegraph*] *Age* and someday you'll be a chief like me"[21] (see figure 9.2).

Such fears were ironic, since messengers were still striking all over the United States. The year 1907 brought an economic crisis, with nonfarm unemployment rising to 16 percent, and messengers struck in Chicago and Omaha, Nebraska. Operators walked out as well: a wildcat strike quickly spread to nearly one hundred cities, involving nearly ten thousand WU and PT commercial operators, with support coming from railway and Associated Press operators as well. This wasn't an official CTU strike, as the CTU "had made no headway worth mentioning in its efforts to organize Western Union employees." But dissent had grown among WU operators over punishments for union involvement, a promised 10 percent pay raise that never materialized, and dissatisfaction with the "sliding scale" (described in chapter 3). The CTU, which had not been prepared for a strike, ran out of money even with donations from the AFL. The strike lasted eighty-nine days, a waiting game that WU and PT won.[22]

The 1907 strike failed, but it motivated Congress to call for an investigation of the telegraph industry. The five-hundred-page report, released in 1909, revealed that after the strike, WU stepped up its efforts to replace skilled, male Morse operators using keys and sounders with (supposedly) unskilled, female Automatic operators using printing typewriters. So-called extra or part-time operator hours had increased, while full-time operator hours decreased. And although no actual blacklist of union operators was found, in practice the duopoly of WU-PT could simply fire all union operators and agree not to hire new operators without (union-free) letters of recommendation.[23]

These aggressive tactics shifted somewhat when in 1909 AT&T, led by Theodore Vail, purchased a controlling stock interest in WU (discussed in chapter 7). AT&T was fighting unionization just as WU was, but under Vail the two companies began to promote the "welfare capitalism" (or "corporate paternalism") tactics of the corporate-backed National Civic Federation. Even though he still fired operators for union activities, Vail began to improve conditions in the telegraph offices, increased WU wages 50 percent between 1910 and 1913, and instituted both a pension fund and a loan program—even the CTU journal hailed his "broad mind and liberal views." The Vail years were brief—in 1914, AT&T gave up its stake in WU under antitrust fears—but Vail's groomed successor Newcomb Carlton continued Vail's reforms. Under Carlton, operators could receive bonuses of up to 7 percent of their annual pay, and even messengers could receive up to a $25 bonus.[24]

Such reforms foreshadowed a new labor strategy for WU: creating an in-house labor organization for its employees that could effectively compete against independent labor unions. In 1914, as Carlton started a friendly new company newsletter, he first considered creating a company union. Additional motivation came with the 1915 report of the Commission on Industrial Relations (CIR), which recommended a government takeover of telegraph. The report declared, "The crux of the whole question of industrial relations is: Shall the workers for the protection of their interests be organized and represented collectively by their chosen delegates, even as the stockholders are represented by their Directors and by the various grades of executive officials and bosses?" The answer for the CIR was yes.[25]

By 1917, only twelve company unions existed in the entire United States, but then the country entered World War I. Nearly 4,500 strikes broke out, involving over 1 million workers, due to the rise in wartime prices without a corresponding increase in wages. At the same time, unemployment fell to less than 2 percent, and factories experienced heavy job turnover as workers moved through many jobs for better wages each time. Looming over it all was the recent Bolshevik Revolution in Russia, precipitating a fear of "Reds" at home. Eager to avoid a general economic upheaval, the U.S. government acted to safeguard key infrastructure industries. In December 1917, the government took control of the railroads. Western Union feared that communications were next in line, and they were right.[26]

But before securing the communications industry, President Wilson appointed a National War Labor Board (NWLB) to settle irreconcilable industrial disputes, with five labor and five management representatives plus former president Taft and the former chairman of the Commission on Industrial Relations, Frank P. Walsh. Wilson said that management should not interfere with or discriminate against unions, but he also prohibited unions from forcing employees to join up. The goal of both policies was to avoid strikes.[27]

Yet striking was exactly what WU and ADT messenger boys in New York City were doing at the time. A messenger strike that began in February 1918 quickly grew to three hundred boys—25 percent of the New York City total at the time. The boys demanded a half-cent piece-wage increase to 3¢ per message, an increase from 15¢ per hour to 20¢ per hour for "temp" work, and a ten-hour day. WU responded by increasing telephone deliveries, by substituting telegraph clerks and linemen for messengers, and even by hiring messenger girls. Though Western Union succeeded in breaking the strike after a few days, the messengers gained both union and government involvement at a critical time and place. Over four hundred messengers attended a meeting at New York City's Yorkville Forum, known as "the Socialist headquarters of the upper east side." The AFL was one of the sponsors of the meeting, and a special representative from the state department of labor attended as well. Western Union executives certainly would have taken notice of this.[28]

Western Union president Carlton attacked such union organizers just a month later, in the March 1918 *Western Union News*: "We know that more than 95 per cent of our employees aim to give the company a square deal. But less than 5 per cent are Bolsheviki who would, if they could, do to the Western Union what has been done to Russia." President Wilson's NWLB responded by ordering WU to cease discriminating against employees who joined unions. But Carlton refused. On June 11, 1918, Wilson sent letters to both WU and PT urging them to comply with the NWLB decision. Postal Telegraph agreed, but WU again refused. Carlton, continuing the WU tradition of trying to appease its employees under company terms, unveiled his company union.[29]

Carlton, like his mentor Vail, thought of WU as a public utility, entitled to receive the same protection against strikes as the post office (even though WU had fought for half a century to remain a private corporation, immune from democratic public control). Western Union first denied that "their employees" made any grievance at all to the NWLB, because by definition if WU employees joined a union like the CTU, they were subject to dismissal and no longer WU employees. Who did file the grievance, then? Carlton stated, "Though there are some 70,000 Morse operators in the United States and Canada, *only 2,500 of them are members of this organization*, and they are chiefly Canadians." According to WU, such "propagandists," far from being WU employees, weren't even Americans.[30]

Next, Carlton argued that WU should be immune from outside unions because of its special status as a "public utility," especially in wartime. Since the post office was allowed by the government to thwart unions, so should WU as both institutions provided a necessary public communications service: "there are two national systems of communication universal in character and open to the public at low cost—The United States Mail and the telegraph system. Each must maintain an uninterrupted, efficient service if confusion and disorder are not to take the place of well-ordered national procedure." But Carlton forgot to mention that WU paid its messengers much less than any postal carriers.[31]

Finally, after all this, Carlton had the audacity to claim, "The Western Union has no quarrel with the principle of unionism." Western Union proposed to hold a secret ballot in which its employees would choose to join either the CTU or a new company union "representative of all employees" (even though only employees over age eighteen would be allowed to vote in such a union, effectively removing the messengers). But this proposal was disingenuous; Carlton said that even if employees chose the CTU, WU wouldn't be obliged to recognize the demands of that organization. Western Union would only promise not to fire employees for belonging to the CTU, though they would still fire any striking employees.[32]

Carlton's argument backfired. The government, agreeing that the telegraph was as crucial a communications system as the post office, took over both the telegraph and the telephone industries in August 1918, putting them under the command of the postmaster general. However, post office control of the wire communications industries wasn't much different from private control. Postmaster Albert Burleson created a Wire Control Board made up of three government officials, but AT&T president Vail was appointed as an advisor to this board. The "operating board" that was created also consisted of four telegraph and telephone executives.[33]

Ironically, WU employees trying to unionize may have been worse off under Burleson than they were under Carlton. Burleson, a U.S. representative from Texas for fourteen years, had already tried to eliminate trade unions from the post office before gaining control of the electric communications industries. Burleson made no secret of his anti-union feelings, stating that he was hostile to labor unions in his 1917 report. During the war, he repealed second-class mailing privileges for socialist and anti-war publications. And according to later

NLRB testimony by Newcomb Carlton, Burleson told him, "I am going to put you in charge of Western Union. And I have only one instruction to give you with respect to labor. Don't let the union sons of bitches get the best of you." In light of the NLRB orders and President Wilson's wishes, Burleson did at least prohibit the firing of employees for union membership. This post office control of WU lasted for exactly a year, as the war ended shortly after the takeover took place.[34]

Though it couldn't save WU from government control, Carlton's company union was created anyway in 1918. The Association of Western Union Employees (AWUE) described itself as "neither a trade union, nor a workers union" but a "voluntary association of individuals in the employ of the Western Union Telegraph Company." A grievance procedure was set up, but employees who were members of the AFL were not eligible to join the AWUE, and strikes were disallowed—the AWUE's journal, *Telegraph World*, advised, "most collective effort is based upon some form of conspiracy instead of being the expression of the spirit of co-operation." Not surprisingly, given statements like that, independent unions denounced the two hundred or so company unions in existence at the time as shams, with Samuel Gompers calling them "a semblance of democracy."[35]

But even though the AWUE was clearly an anti-union organization, it signed up nearly half of WU's forty thousand employees during its first year. A June 1919 CTU operators strike was held in vain—the AFL union could not break the AWUE. The CTU was heartbroken, its president resigning in shame. The CTU's failure mirrored that of the AFL at large—with the "Red Scare" in full force after the end of the war, kicked off by the notorious "Palmer Raids," the AFL began to purge itself of "radicals" in the 1920s, and its membership shrank from about 5 million to only 3 million in the first few years of the decade. On the other hand, company unions prospered in the 1920s. By 1922, there were 726 company unions among 385 firms, and though many soon failed, the number of company unions remained around 400 through the 1920s. WU's AWUE continued to grow through the 1920s until it commanded nearly 100 percent of WU employees by 1930—advertising itself to members as "Your Friend—Counselor and Advisor."[36]

Ironically, the AWUE, born in the midst of a widely publicized messenger strike, didn't cater at all to messengers at first. When WU demonstrated the AWUE's effectiveness by distributing over $3 million to employees in a new profit-sharing plan in 1920, this money was divided among only 28,676 of the 60,500 total WU employees—messengers and temporary workers were excluded. AWUE rules allowed any employee aged eighteen or above who had been with WU for over three months to join, thus excluding all but the oldest messengers on the basis of age and job tenure. During the AWUE's first year, not even a

New York City strike of nearly two thousand WU boys (so serious that state mediators were called in) bothered the new union.[37]

Though the AWUE continued to ignore the messengers, as the number of messengers increased in WU and PT through the telegraph heyday of the 1920s, the CTU finally noticed them. It might seem obvious that messengers would be included in a telegraph union that already counted among its members "chiefs, supervisors, Morse operators, automatic operators, telephone operators, all the clerical workers, linemen, gangmen, foremen, cooks, water-boys, installers, instrument repairmen, tailors and janitors." But messengers had been thought of as subcontracted, temporary, child-labor employees for so long that bringing them into the adult union was a big step.[38]

In the end, it was their sheer numbers and their usefulness in strikes that sold the messengers to the CTU. In 1925, the Canadian National Telegraph, one of the CTU's strongholds, employed three hundred messengers. Proponents of messenger membership argued that these messengers had been an "immense benefit" during a recent Canadian Press strike, but now "the management of the Canadian National Telegraphs have foreseen that potential strength, and by the equipment of gyms, drills[,] educational lectures, telegraph schools, and other means are breaking this potential field to their own use." Among all companies where the CTU was active, proponents estimated that the union could add three thousand messengers—not including the advantage of messenger numbers in any future WU organizing drive. One delegate, a former messenger himself, argued on behalf of the boys by pointing out, "These messenger boys will have an opportunity to learn secrets. They will be witnesses of various affairs, and their information and assistance in any strike will be very valuable." Thus the delegates finally voted in favor of admitting the messengers, understanding them not as boys, but as "future telegraphers."[39]

The prosperity of the 1920s ended for much of the nation in 1929, with the Great Depression bringing a change in labor markets and a rise in union activity. Boys with jobs held on to them into their early adult years, and men who ordinarily wouldn't think to apply for "boys' jobs" sought out such positions in earnest. Thus the average age in the messenger service began to rise in the early 1930s, causing messenger work to be viewed differently by employers, unions, and government regulators. This change came just as the AWUE had signed up nearly 100 percent of its adult target group, some 36,603 adult employees. But the independent union threat was returning—overall union membership would rise from 3 million to 9 million in the 1930s. Thus in April 1934, at a time when the AWUE counted only five hundred messenger members nationwide, the company union began to organize messengers at the request of WU itself, "confined to certain divisional offices where disturbing influences might arise," especially when new messengers demanded cancellation of their uniform rentals.[40]

The AWUE had for some time allowed messengers to become "associate" members by paying dues;—however, associate members lacked voting rights. By June 1934, AWUE bylaws allowed all messengers to become voting members, and the organization proclaimed, "thousands of Messengers are affiliating with our Organization throughout the United States." In November 1934, the AWUE claimed that it had signed up three thousand of the estimated twenty thousand WU messengers across the United States, imploring to its members, "Let's make it 100% strong—Every messenger a member!"[41]

But what did the AWUE do on behalf of its new messenger members? The AWUE described messenger areas within company offices as lush, in spite of the fact that, as seen in chapter 4, these spaces themselves had been increasingly hidden from consumers. In Atlanta, for example:

> Lounging rooms have been provided that rival those in the finest theaters. Large marble shower baths, lunch room, the very latest equipment for keeping the uniforms in first class condition, including one room especially equipped for drying out uniforms on rainy days, a large electric dryer doing the work, and numerous other details, every one adding to the comfort of the boys.

Such corporate paternalism amenities were yet another attempt to carve out messenger spaces within telegraph offices to better control the boys. Even with "lounging rooms" and "baths," messengers were rarely paid for time spent waiting for messages and changing into uniforms.[42]

Furthermore, even the most comfortable messenger spaces were company innovations, not union gains, since they had been installed before messengers even became AWUE members. The AWUE often echoed official WU company policy in this way. When the NRA proposed a "Code of Fair Competition for the Telegraph Communication Industry" in 1934, including a minimum wage for messengers of $10 per week, the AWUE translated the rate into 25¢ per hour, and said that messengers would be paid only half that rate for the time they sat on the bench. Later the AWUE argued that since WU's twenty thousand messengers earned an average wage of $6.50 per week, raising that average to $10 would cost $3.64 million per year, resulting in messenger layoffs. Thus the AWUE actually helped WU preserve its low-paid, child-labor messenger force, even while appearing to act in the messengers' best interests. In the end, the AWUE was still a company union, and WU was still a company hostile to unions.[43]

Ironically, it was the messengers who struck the first blows against the AWUE. First, in March 1937 the Toledo WU messengers affiliated with the

CTU and went on strike. The WU division president, in a move similar to that of ADT management decades before, sent telegrams to the parents of the messengers warning them of the outside union and encouraging their sons to return to work under the AWUE instead. The strike was broken, but it had an important consequence behind the scenes. Western Union vice president J. C. Willever, fearing efforts of "outsiders to stir up discontent among our messengers," wrote to all AWUE division heads: "I suggest that you keep your ear close to the ground and arrange to get the Association in promptly on any sign of dissatisfaction or unrest." Thus WU documented the fact that it exerted considerable control over AWUE policies at the highest levels.[44]

Close on the heels of the failed Toledo strike, a Seattle messenger strike in June 1937 closed that city's main Western Union office and all nineteen branch offices when 116 boys demanded higher wages. Some four hundred workers were idled by the closings, but the effects didn't end there. The messengers were again affiliated with a CTU local, and they filed unfair labor practice charges against WU on July 26. Western Union responded that it recognized only AWUE messengers as a proper bargaining unit, not CTU messengers.[45]

This was precisely the kind of situation that the National Labor Relations Board had been set up to solve. Part of President Franklin D. Roosevelt's New Deal reforms, the NLRB had been authorized one month after FDR's first anti-Depression measure, the National Recovery Administration, was declared unconstitutional by the Supreme Court. The NLRB was meant to enforce the National Labor Relations Act (or Wagner Act), which defined a list of "unfair labor practices," including financing company unions, employing company spies, firing employees for organizing, and refusing to bargain with unions. The act would be challenged, but was upheld in the Supreme Court two years later.[46]

The Seattle messenger strike was the first time WU practices were scrutinized by the NLRB. Newcomb Carlton himself, by now no longer WU president but chairman of the board, traveled from New York City to testify at the hearing. But the NLRB sided with the messengers union. It ordered the Seattle WU office to cease encouraging membership in the AWUE and to recognize the CTU instead. The messengers had finally shown themselves to be important to telegraph unionization, partly because of their willingness to act and partly because of their sheer numbers, especially in urban offices. Wrote one WU manager later, "It was inevitable . . . that the backbone of the unionizing movement would be the messenger, the man from below."[47]

The union that learned the most from these events was neither the CTU nor the AWUE, but a new telegraph union formed in the 1930s, the American Communications Association (ACA). The ACA had gotten its start among a particular subset of telegraphy workers: wireless telegraph operators. Originally organized as the American Radio Telegraphists' Association (ARTA) in 1931, it soon developed a broader agenda of uniting all communications workers. In 1937, it affiliated with the Committee for Industrial Organization (CIO), expanded its target to land telegraphs, and began to organize New York Postal Telegraph employees.[48]

In affiliating with the CIO, the ARTA was doing more than simply finding a national sponsor—it was consciously rejecting affiliation with the CTU and that union's parent organization, the AFL. In 1935, John L. Lewis, after working first as an organizer for the AFL, then as a member of the United Mine Workers, and finally as the United Auto Workers president, broke with the AFL and formed the Committee for Industrial Organization. While the AFL stood for conservative, careful organization of craft and skilled workers, the CIO advocated mass organization of as many industrial workers, skilled and unskilled, as possible. Thus the AFL viewed the CIO as a threat—a "dual union."[49]

The new CIO unions often handled grievances on the shop floor with "brass knuckles unionism," where organizers themselves held "mini-sit-downs" or "quickies" to battle individual foremen. The CIO was overtly political, supporting Roosevelt and the Democrats in their push for public welfare benefits. And the CIO more readily admitted "radicals" like socialists and Soviet-sympathetic Communists. AFL unions, on the other hand, usually handled grievances using paid business agents. They did not endorse political parties, preferring to provide their own welfare services. And they were overtly "anti-Red," accusing their enemies of Communist subversion. By 1937, the year the ARTA joined the CIO, the CIO counted 3.7 million members after victories in two formerly open-shop industries, auto makers and steel makers. This was at a time when the venerable AFL counted only 3.4 million members of its own. Together, the unions had a formidable constituency, but the two groups were unable to agree to merge. The CIO's break from the AFL was made official in 1938 with its renaming as the Congress of Industrial Organizations.[50]

The CIO was the right place for the ARTA, because after less than a year, the upstart telegraph union had organized the second-largest U.S. telegraph company, Postal Telegraph, ousting the CTU in the process. After it won, the ARTA changed its name to the American Communications Association, reflecting its enlarged mandate. All the CTU could do in response was to accuse the ACA of communism.[51]

The CTU knew that the ACA would now set its sights on the same prize the CTU had been chasing for years: Western Union. But the ACA had learned both from the past mistakes of the CTU and from the past successes of the ephemeral messenger unions. In its first move against WU, the ACA filed unfair labor practices charges with the National Labor Relations Board, charging WU with discrimination against the ACA and in favor of the AWUE—just as the Seattle messengers union had done successfully a few years before.[52]

The ACA used the AWUE's haphazard approach to organizing the messengers as part of its evidence that the AWUE was a company union, demonstrating that the AWUE tried to organize messengers only when ordered to by WU managers—citing J. C. Willever's order to the AWUE after the 1937 messenger strikes. For example, in response to this order by WU, the president of the AWUE Gulf Division had told his local officers to "Select from six to twelve boys of the highest type; high school graduates, clean cut, intelligent, good personality" and to "Sell these boys on the fact that the A.W.U.E. is the parent organization that deals for all employees in all matters pertaining to the welfare of Western Union employees." He also warned, "Mass meetings of all the boys must be avoided as only demands and more demands will be made upon you." It seemed clear from such statements that the telegraph company's "association" was as far from a "union" as could be.[53]

The ACA pushed messenger issues not only in the courts, but at the bargaining table as well, so much so that *Business Week* mistakenly referred to them as a "messengers' union." In October 1938, the Fair Labor Standards Act, introduced by FDR a year earlier, instituted a 25¢-per-hour minimum wage, rising to 40¢-per-hour in seven years. Western Union immediately objected, seeking a special exemption from Congress, saying that it could not possibly afford to pay their messengers that rate. But the ACA fought for the messenger minimum, secretly delighted that Western Union disregarded the law: "I think a hearing on the messengers, provided we can pin it on WU, would give us a chance for some publicity and make the messengers good and restless everywhere," wrote one organizer to the ACA leaders. ACA locals wrote direct appeals to messengers in large cities: "*Only* the ACA has a program and is making a stiff fight to win the $11 [weekly] minimum for messengers, and to defeat any merger except on terms which will fully protect the jobs, wages and working conditions of all employees." When the Senate Committee on Interstate Commerce held hearings on whether or not to investigate the telegraph industry, the ACA was there, not the CTU, and the ACA brought messengers to testify on their own behalf.[54]

Though the CTU publicly proclaimed its support for the messengers as well, in private it attacked the ACA for courting this volatile group of workers. In 1938, the CTU president wrote to the CIO's John L. Lewis:

ACA is built upon the doubtful foundation of terribly underpaid messengers and other Postal employees who face the probability of being thrown out of employment by liquidation of their bankrupt employer [through merger of Postal with WU], and therefore are desperate and willing to do whatever they believe can give them temporary security. . . .

We can organize Western Union if CIO does not add further confusion to the situation and play into the hands of Western Union Telegraph Co. by furnishing another $100,000 or more for another carnival of ACA activities with messenger strikes, slowdowns and other forms of sabotage which only helps drive telegraphic business to the Bell Telephone System.

Here the CTU revealed its AFL craft-union biases. "Desperate" and "underpaid" employees did not make for strong unions, nor did employees who engaged in "carnival" strike actions. But these were the very employees that were swelling the ranks of the CIO.[55]

Using the messengers made sense for the ACA. For example, they began to organize San Francisco WU messengers in May 1939, and by September they were ready to strike. With the messengers out, the San Francisco office was closed down. Nonmessenger ACA members joined their young colleagues a few days later. The strike was considered serious enough that the secretary of labor intervened, calling WU president Roy White and ACA president Melvyn Rathborne to Washington to settle in November 1939. Western Union boycotted the meeting, again arguing that the AWUE was the only union they recognized, but a month later they accepted the secretary of labor's recommendation that the strike end with all employees returning to their jobs—and with the ACA as the new union in the San Francisco office.[56]

The ACA's use of the messenger boys against Western Union's AWUE worked. By late 1939, the AWUE admitted that "An increasing amount of conference time is being utilized in the solution of messenger working condition problems," revealing the novelty of messengers being factored into such decisions at all. That year, the NLRB trial examiner ruled that the AWUE was dominated by WU, and could no longer be recognized as the bargaining agent of the employees. Western Union appealed, especially concerning the question of whether the AWUE only tried to organize messengers to keep WU strike-free, as company managers realized "the Board lays particular stress on this evidence." The AWUE now argued that organizing messengers had never made sense in the first place, as such employees didn't deserve a union vote when they only paid dues once and then quit a month later. But the appeal failed, and in June 1940 the AWUE's *Telegraph World* folded. By using the messenger boys as a point of leverage, the ACA had broken WU's twenty-year-old company union.[57]

The ACA could not claim victory, however. The passing of the AWUE left a power vacuum, with nearly all of the nation's WU employees without union representation. Now the battle between the ACA and the CTU would intensify because of what everyone knew was on the horizon: a government-ordered merger between the ailing but ACA-dominated Postal Telegraph and the powerful but unionless Western Union. One company would result, and only one union could represent its workers.

The CTU may have been on the defensive, but backed by the AFL it was certainly not weak. And it had not been idle over the preceding years—by 1939, it had organized the two large commercial telegraph systems in Canada, the four press agencies in the United States and Canada (including the formerly anti-union Associated Press), the "leased-wire" financial houses, and the marine radio telegraphers. What the CTU needed to break WU now, it argued to the AFL, was more money: "just as soon as the NLRB disestablishes the company union, we will be confronted with requests from all over the nation for affiliation with CTU far beyond our present financial capacity to cope with."[58]

But the CTU was frightened, and eschewing the spirit of union brotherhood, its main attack on the ACA was ideological: "We possess documentary evidence to prove conclusively that ACA is officered and controlled by Communists, or Communist affiliates, and that money has been given them freely [$100,000 by the CIO] to further the much-desired control of communications." With World War II threatening to envelop the United States, the CTU argued that telegrams handled by ACA shops were not safe: "*Telegrams addressed to or signed by AFofL officers* are not safe from perusal of *Communist leaders* when transmitted via *POSTAL, RCA* or *Mackay*."[59]

The ACA *was* more radical than the CTU in that it worked harder for the low-paid, low-status messengers. The ACA filed a complaint against WU in July 1941 with the Department of Labor, arguing that if messengers were to supply their own bicycles, then at least the company should pay the upkeep costs. It was a seemingly small request, but it affected every single bicycle messenger, and it was a winnable goal. The Department of Labor would rule in the ACA's favor a little over six months later.[60]

The ACA's messenger focus had real consequences. By 1940, elections for telegraph union representation had been held in most major cities, and a pattern was set that would last for the next decade. The ACA and the CTU were forced to share bargaining power within Western Union, but unevenly: CTU power was spread out among the states, and ACA power was concentrated in New York City. As the nation's center of telegraphic communication and the site of WU headquarters, New York City in 1940 employed over two thousand of the industry's fourteen thousand messengers, more than anywhere else.[61]

Having captured New York City, the ACA continued to press for the messengers in the other large urban telegraph markets. When four hundred Chicago messengers struck in May 1941, a settlement was negotiated with ACA mediators, including seniority assurances, a lunch period after four hours of duty, no more fines for tardiness, a company promise to keep uniforms in good condition, and an "elimination of speed-up." The election held in July 1941 to decide which union would represent the one thousand Detroit WU employees was won by ACA in part by the votes of 200 eligible messengers, even though 150 other messengers were declared ineligible to vote because they had started with WU only a month earlier. The ACA bragged that the CTU couldn't match its messenger organizing tactics: "The CTU-AFL stooped to a new low when they tried to capture the 200 eligible messenger votes by feeding whiskey to 17 year old messengers and by passing out 'courtesy police cards' which were supposed to grant immunity to the bearers from prosecution by the police for traffic violations." But more important, the ACA followed through in its courting of the Detroit messengers by remembering them in the new contract with WU, proposing free bicycle maintenance, full "tours" (shifts) with lunch hours, and time and a half for the big telegram holidays like Christmas.[62]

Thus, when World War II came the landscape of telegraph unionism was uneven and contested. The ACA had control of WU's main competitor, PT, but that company had been in trouble for some time. The CTU still held on to Canadian press and brokerage operators, but was running low on funds. In large urban markets where messenger numbers made a difference (like New York City), the more radical ACA had a foothold in Western Union; but in the rest of the nation, the more conservative CTU was more readily chosen by a workforce accustomed to twenty years of company unionism. Finally, Congress pushed the long-debated telegraph consolidation forward, fearing that disarray in WU and PT would endanger the efforts of the military. Only one union could prevail.

Suddenly the ACA was at a disadvantage, in part because of its reliance on the messengers. As the government investigated conditions in both WU and PT in preparation for the merger, familiar testimony about low messenger wages, long messenger hours, costly uniform fees, and harsh rules was read before Congress to denouce current telegraph practices. But the government's main fear concerning the merger was mass layoffs, not wages or working conditions. In this context, the high endemic messenger turnover rate was a hidden asset to the merger proponents; messengers, the biggest single category of employees, could be expected to quit with such regularity (especially during a wartime industrial boom when higher-paying jobs were plentiful) that they represented no layoff concern at all. ACA arguments about the plight of the messengers fell on deaf ears.[63]

Finally in 1943, well into the war, Congress amended the Communications Act of 1934 to provide for a merger of WU and PT. New union elections were necessary for all 50,000 telegraph employees, including the 12,000 messengers: PT's 9,000 employees were still covered by an ACA labor contract, as were 6,000 WU employees in the New York City Metropolitan Division, but 31,000 WU employees nationwide were represented by the AFL. The ACA and CTU would have to compete all over again for merged company representation, and each union wanted different spatial and temporal parameters for the election.

The CTU pushed for an immediate election to pick a single bargaining unit for all of WU, threatening a strike if these demands weren't met; the ACA pushed for a delayed election and for separate bargaining in each of the 105 divisions of WU. The NLRB offered a compromise: an election would be held in January 1945, and the sixty thousand WU employees would be divided into six separate geographical bargaining units.[64]

Although the ACA was still more aggressive than the CTU in targeting messengers in the new organizing campaign, the boys were losing their appeal for the so called messengers union. In August 1944, ACA president Joseph Selly urged the NLRB to "eliminate from the election those who came on the payroll after October 8 [1943]," because between the date of the merger and the date of the election, "approximately 5,000 messengers" had "gone out of the industry in the course of the usual turnover." In other words, even though on the day of the merger the ACA represented fifteen thousand employees, on the day of the election they only represented some six thousand employees. Even though the ACA argued publicly that increasing messenger wages would directly "improve the delivery service rendered the public by the company," behind the scenes new ACA organizers were warned, "work should be concentrated among non-messenger employees . . . since the turnover among messengers is so great that the number eligible to vote (that is, on the payroll long enough) will be negligible."[65]

Ironically, just as the ACA stopped courting messengers, the CTU stepped up its messenger-recruiting efforts, since in 1943 only five hundred of the CTU's twenty thousand members were messengers (contributing less than 1 percent of the union's dues). This low membership resulted in an uneven pattern of messenger contract provisions in the various CTU locals across the country. In 1942, Kansas City and Cleveland prohibited making messengers do clerical work at messenger pay rates; Cleveland and Washington declared that messengers would be considered for promotions; Kansas City said bicycle and foot messengers would have preference for promotion to motor messengers; Kansas City, Cleveland, and Washington required three days' notice before messenger layoffs; and Washington prohibited lockouts, strikes, or work stoppages for messengers. While some CTU locals might have been interested in the messengers, the national organization was unable to mobilize the boys.[66]

The NLRB-sponsored election was finally held in January 1945, and the geography of labor had an important effect on the results. Although the nationwide vote was roughly twenty thousand to eleven thousand in favor of the CTU, the ACA prevailed in one of the six divisions: New York City. This ACA victory was likely due in part to the still-substantial messenger numbers in New York City. But telegraphy was made up of more than just urban messengers, and in the end the CTU controlled around fifty thousand of the sixty thousand WU workers. The ACA controlled only eight thousand.[67]

Ironically, when the ACA and the CTU finally negotiated their separate contracts with WU, the CTU affiliates nationwide ended up benefiting from the ACA messenger gains made in New York City. Foot and bicycle messengers earned the lowest wages of all WU employees, and both the CTU and the ACA were trying to raise this wage floor. But for the CTU, fighting for the messengers involved asserting that those messengers were no longer boys, but men: "The majority of messengers in Western Union today are not school children; they are breadwinners in every sense of the word, and in common decency are entitled to a living wage." They argued that raising the messenger minimum wage to 65¢ would help stabilize that job category as "the proper stepping stone to the higher skilled job classifications." In contrast, the ACA still recognized that messenger boys *were* boys, and wasn't shy about the fact (especially in published photos). The ACA simply demanded that "all messengers . . . receive the full benefits of the contract." WU's own position was closer to that of the ACA: messengers *were* children in the eyes of the company, mostly under age eighteen, and deserved to be treated as such—with no 65¢ minimum.[68]

With wage demands linked to a higher messenger minimum, both the CTU and the ACA contract negotiations with WU initially broke down, and were referred to government arbitration—but at two different scales of government. The ACA wage ruling, handed down by the New York Regional War Labor Board, turned out to be more generous than the AFL wage ruling, handed down by the National War Labor Board. To resolve the disparity, the National War Labor Board finally ruled that WU employees would get a wage increase in line with (but slightly less than) the ACA contract demands that had been approved earlier by the New York Regional War Labor Board. Thus CTU employees nationwide benefited from their ACA competitors in New York City.[69]

In practical terms, the CTU had all but won the company; nevertheless, in New York City, worldwide center of telegraph communication as well as home to the largest single group of telegraph messengers in the United States, the CTU was forced to concede. Yet for the messengers, a large battle was lost. Once their wages were brought in line with national legal minimums, WU opted to increase mechanization and subcontract out to post office and taxi services to carry its telegrams instead of keeping a large messenger force of its own. The success of the unions (and the government) in making of messengers into "men" instead of boys thus marked the end of the messengers' tale.[70]

CHAPTER TEN

The Telegram Is Dead; Long Live the Messenger

Taking a sociological point of view, messenger service of the past was based on the mass availability of low cost labor. Messengers earned pennies an hour, and worked hard for these pennies. I don't think those who regret the passing of the service want to return to the conditions that made it possible.

—WU president Russell McFall, 1971[1]

I am a bike messenger: a lackey, a laborer, a punk-ass kid who calls this place his home. I am part of this machine, and I am also its observer. I am in the mix most of the time, hardly a self at all, but then I get these flashes, these pictures that assemble not about me but about the city.

—Chicago bicycle courier, 2001[2]

This book has presented the story of the telegraph in a new way. Other historians have already focused on the system builders who invented the wire communications technologies and incorporated them into profit-making enterprises. Other geographers have already traced the changes that the resulting telegraph network wrought on the distribution of cities and the speed of business. But this book has considered the history of a group of "system maintainers" at the lowest level of the telegraph hierarchy, and the lived geography of the telegraph network that both resulted from and set limits upon their day-to-day labor.

These system maintainers were the telegraph messengers, and contrary to their portrayal in previous accounts, these workers were more than just a quaint anomaly in the history of modern American business and technology. Messenger boys were crucial to the functioning of the telegraph network for nearly one hundred years, both in the network's core business of handling information, and in its accessory business of hiring out temporary workers. But more than this, messengers became part and parcel of the key telegraphic product, the telegram, not only advertising it to the world, but carrying with it an aura of importance and urgency that kept it distinct from the products of the competing information networks, the slow letter and the hurried telephone call.

Because of the importance of messengers, the telegraph companies were forced to act to preserve this labor force. They attempted messenger discipline through the mandating of military uniforms, increased messenger speed through the dissemination of bicycles, and claimed to advance messenger careers through the creation of a continuation school. But without real improvements in wages, conditions, and future prospects, nothing could keep the messengers from acting on their own, whether through their regular practice of leaving and reentering the telegraph system to sustain a phenomenal turnover rate, or through their rarer but more sensational ability to strike en masse, bringing a city's information network (for a few days, at least) to its knees.

Yet telegraph messenger boys represent not only a romantic tale of underdogs living at the very bottom of a single corporate pyramid, but also a case study of the workers most obviously caught between the boundaries of the multi-institutional communications internetwork itself. The telegraph companies both phoned messages and mailed them, but could not shed the messenger boys until union demands made them as expensive as adults. The post office did everything it could to distance its letter carriers from the lowly messengers, yet the two groups served nearly identical functions. And even the telephone companies relied on messengers during that critical period when their network was still not dense enough to claim "universal service" alone. Thus messengers functioned as key "boundary workers" in the nation's first electromechanical communications internetwork.

After World War II, the histories of Western Union, its former subcontractor ADT, and the messenger boys themselves forever diverged: the monopoly national telegraph company focused on regaining profitability through "modernization," the district telegraph company focused solely on alarm services, and telegraph messengers everywhere were gradually and silently phased out. This chapter brings the story of the messengers up-to-date by briefly describing how ADT survived as a global security services firm, how Western Union perished through bankruptcy and sell-off, and how the strikingly familiar job of "bicycle courier" has once again emerged in many U.S. cities as a necessary, low-tech complement to the nation's latest form of high-tech telecommunication, the Internet.

One of the ironies of the history of telegraphy is the shift in status between American District Telegraph and Western Union. ADT, originally the local subcontractor of messenger labor to the national telegraph giant, is today a multinational company still in the electrical alarm business. But WU, never able to recover its technological momentum after World War II, has gone bankrupt, forced to sell its very name to avoid public embarrassment.

Although ADT officially left the messenger business in 1924 when it sold its "messenger plant" to WU for $625,000, it still had strong links to the telegraph monopoly. After all, WU still owned 62 percent of ADT's stock. Nevertheless, ADT was finally free to concentrate on what it now thought of as its core business, leasing automatic alarm services to homes and businesses. Over a quarter century later, WU finally sold its remaining stock in ADT to the Grinnell Corporation (formerly known as General Fire Extinguisher) for $12.5 million in a bid to raise cash during a period of postwar losses, severing all ties with its former subcontractor.[3]

ADT would soon stand on its own, after a 1961 Justice Department antitrust suit resulted in a divestiture decision by the Supreme Court some years later. By 1968, ADT had become an independent, publicly traded company. But this would last only two decades; in 1987, ADT was acquired by the British-based Hawley Group Ltd. Hawley eventually changed its name to ADT Ltd., at a time when its U.S. subsidiary, ADT Security Systems, employed twelve thousand people and counted over 25 million clients. Thus an idea for an urban home security network, spawned over one hundred years before in the mind of call-box inventor Edward Calahan, continued to carry this company into the twenty-first century.[4]

Western Union, on the other hand, was not so fortunate in the years after World War II. Though wartime telegraph traffic had kept the company afloat through the 1940s, it also resulted in a 25 percent excise tax on consumer telegrams, an end to messenger-based extras like Dramatized Delivery and singing telegrams, and a return of the public perception of the telegram as a death message. According to one economic historian's assessment in 1947, "A plant that is largely obsolescent, a rate structure requiring thorough overhauling, deteriorated quality of service, high unit labor costs, failure to withstand severe competitive pressures—these are some of the ills that currently beset Western Union." But WU was free from competition with PT, and had weathered the worst from the union battles between the CTU and the ACA. Thus WU faced midcentury with an uncertain future.[5]

Starting in 1946, the company began to respond to its critics. It reinstated its messenger delivery services and spent $60 million buying new "reperforator" machines, pushing its FAX system, building a new microwave-based "radio beam telegraph network," and battling AT&T over rights to teletype patents and leased-line customers. Nevertheless, WU lost $11 million in the first year after World War II, and although the company showed a slight profit due to a two-month telephone strike in 1947, it lost over $4 million each year in 1948 and 1949. The revived messenger errand service helped, bringing in $1 million of revenue in 1948, but it was not enough to stem the losses. When new WU president Walter P. Marshall took over in 1948, he knew that WU had to change or perish.[6]

Marshall's "modernization effort" involved geographical as well as technological restructuring. Western Union sold its own real estate, including its signature New York City headquarters building at 60 Hudson Street (going for $12.5 million), and then leased this office space back when needed, saving $2 million per year. The company reorganized its own space by splitting national operations up into regions that focused less on New York City and other large urban centers, and more on new suburban markets. And after realizing that an elite group of only 142 offices (less than 2 percent of all offices) handled fully 75 percent of all its message traffic, WU closed more sites around the nation, contracting with 22,000 existing merchants such as drug stores and hotels to accept telegrams on commission.[7]

The map of telegraph labor underwent restructuring as well. After the labor unrest of 1946, many legislators decided that the 1935 Wagner Act had given labor too much power, and Congress passed (over president Truman's veto) the Labor-Management Relations Act of 1947, known as the Taft-Hartley Act. This set of amendments to the Wagner Act severely limited union power, banning such tactics as sympathetic strikes, secondary boycotts, and mass picketing. It also required unions to renounce the Communist Party (legitimizing Red-baiting) and made unions legally liable for damages if members violated contracts (especially if union workers struck). Finally, the act allowed states to pass open-shop or right-to-work laws that weakened labor's power to set up collective bargaining units. Of the two big labor organizations, the AFL, while not necessarily supporting the Taft-Hartley Act, was more accomodating to it than was the CIO.[8]

This was not surprising. The refrain of the AFL-affiliated CTU had been for many years that the CIO-affiliated ACA was a Communist union, and as World War II ended and the Cold War began, the ACA's own parent organization began a purge of alleged "Reds" from its ranks. Then came 1949, a high-water mark for fears of Communism: the Soviet Union exploded its own atomic bomb and the Communist revolution in China was completed. The CIO president publicly attacked all remaining pro-Soviet members, and while it was difficult to prove that particular persons were "card-carrying Communists," it became easy enough to expel unions whose records seemed to indicate dangerous positions on such issues as the Nazi-Soviet nonaggression pact and the

Marshall Plan. The ACA was one of the eleven CIO unions targeted in this way as "pro-Soviet."[9]

In 1950, the same year Senator Joseph McCarthy began his anti-Communist hearings, ACA executives testified before the CIO on charges of Communist support. Even though they fought against what they called "The Plot (. . . That Failed) To Destroy Our Union," the ACA leaders were ousted soon after, a move that WU surely welcomed—president Walter P. Marshall himself warned the ninety-three New York City WU messenger school graduates in the class of 1951, "Communists are aiming much of their propaganda at young people like yourselves." That same year, the government held its own hearings on whether the alleged Communist control over the ACA had endangered the nation's telegraph industry during the war. J. L. Wilcox, WU vice president of employee relations, testified that WU relations with the ACA were cordial until November 1945, when "the Soviet Union and the United States started parting company." What was the first instance of danger that Wilcox cited, the first evidence that some five hundred ACA shop stewards in New York City were directly "under the discipline" of Communist officals? A strike of 140 Brooklyn automobile messengers on Thanksgiving 1945, a heavy telegram day for WU.[10]

By the 1950s, the union battles in telegraphy had reached a truce. The post-purge ACA was no longer a threat to the CTU, and in any case the AFL and the CIO would themselves merge in 1955. Even though wage levels of operators, clerks, and messengers had doubled between 1941 and 1946, with labor costs representing some 80 percent of WU expenses in 1950, the vast geography of Western Union, encompassing the many union-hostile right-to-work states created after the Taft-Hartley Act, helped keep labor's demands in check. The last big strike came in 1952, when the CTU, demanding a 50¢-per-hour wage increase and a forty-hour week, struck for fifty-three days, costing WU $7 million. *Barron's* later said that the strike helped purge the "pent-up ill-will of decades." For the first time since their company union was disbanded in 1939, WU finally had a measure of control over its labor force.[11]

Western Union used this control to eliminate as many strike-prone employees as possible—which by this time meant both skilled operators and unskilled messengers. After all, the decade from 1940 to 1950 had seen a massive increase in messenger wages, from a child's wage of 20¢ per hour to an adult's (minimum) wage of 75¢ per hour (a 275 percent increase). During the same period, average operator wages rose from 50¢ per hour to around $1.25 per hour (a 150 percent increase). In 1949, "Western Union's new miniature facsimile machine" was finally installed in a customer's office in New York City for daily use; Marshall proclaimed, "It saves the time of picking up and delivering telegrams by messenger, and reduces the sending of a telegram to the simple process of placing the message on the cylinder of the machine and pressing a button." Soon automated "messenger stations" began to appear in cities, churning out batches of telegrams that were later collected by a motor messenger and then delivered by car or "Telecycle" (motorcycle), removing the need for foot and bicycle messenger boys. But perhaps the most dramatic change was the growth in leased circuits—the private lines that customers rented from Western Union for use with their own telegraph printers and FAX machines. Receipts from such arrangements shot up from only 2 percent of all WU revenue in 1947 to around 14 percent of all WU revenue by 1957. By that same year, WU kept active less than half as many call-boxes (about 112,000) as it had just a decade before (232,000).[12]

In 1950, after WU cut $6 million in labor costs and reaped increased telegram revenue from the outbreak of the Korean War, the company finally showed a profit of $7.3 million. But the company was still fighting messenger minimum wages at midcentury, again seeking special exemption to keep its foot and bicycle messengers at 65¢ per hour instead of raising them to the new 75¢-per-hour minimum (at a time when 70 percent of all messengers made less than $1,500 per year, and 29 percent of all messengers made less than $500 per year). Yet messengers were now only picking up 25 percent of all telegrams sent (down from 50 percent in 1930), and only delivering 50 percent of all telegrams received (down from 75 percent in 1930). Thus Western Union celebrated its centennial in 1952 by advertising "the terminal handling system of the future," public "tele-fax" facsimile machines, which WU named "the magic messenger" and "the miracle messenger." By 1960, WU's nonmagical human messenger force had dropped from over 12,500 to around 7,000, with an ever increasing percentage of those messengers working part-time only. Even motor messengers were declining, as local offices subcontracted delivery and pick-up duties to local taxi companies.[13]

Even though WU recast its messengers as quaint relics from the past, it may have been the very availability of cheap messenger labor in the telegraph industry that helped to delay technological change in the first place. After all, during the 1920s and 1930s WU resisted the adoption of all of the electro-mechanical telegraph innovations that its rival AT&T was pioneering—carrier circuits, teletypewriter exchanges, and automatic switching among the most important. Western Union's so-called "modernization" began in earnest only after child labor laws, compulsory education laws, minimum wage laws, and messenger union membership all combined to finally make messenger boys no more inexpensive than messenger men. All of this suggests that a kind of "technological momentum," involving reliance on messengers, was at work. Western Union managers, familiar with a sociotechnological system of "final delivery" based on boys and bikes, uniforms and piece wages, worked to keep this system going and growing as a way of avoiding disruption and risk, both to their core

Figure 10.1 *Western Union employees by percent, 1919–1947*

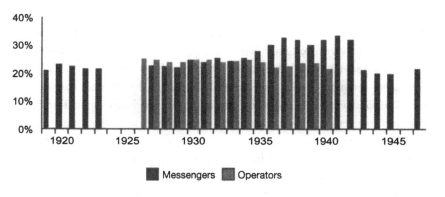

Messengers ■ Operators

Sources: U.S. Senate (1941); WU, *Annual Reports* (1971–47); U.S. Communications Policy Board (1951); WU, "Exhibits originally presented to the President's Communications Policy Board in 1950, . . . extended, revised, or otherwise brought up to date" (1954), WUA, 1993 addendum, series V.

business of telegram handling, and to their personal positions in a labor-intensive managerial hierarchy.[14]

This hypothesis is supported by employment trends from the 1920s through the 1950s. After all, WU continued to make high profits during the 1920s while maintaining a messenger force that comprised over 20 percent of its entire workforce—roughly the same number of messengers as telegraph operators. Through the 1930s, as WU fought both the Depression and the competition from AT&T and the post office, the number of messengers fluctuated a bit, but never dropped below ten thousand. More important, the *proportion* of messengers out of all telegraph employees actually increased in the late 1930s, until messengers made up one-third of all WU workers by the time World War II began[15] (see figure 10.1).

Two trends help to explain this reliance on messenger boys. One was the long-term effort by WU to close offices, saving both rent and operator wages. Even though WU would later claim that its office closings were due to a broader postwar modernization effort, the trend began during the Depression in the mid-1930s. The total number of WU offices in the United States had held relatively constant at around 25,000 throughout the 1920s, but after the Depression the company had closed over 5,000 offices by the start of World War II—mostly small, rural railway and agency offices. This meant that more messengers were needed to cover the territory lost through these office closings. When WU and PT merged in 1943, the number of new WU offices in the combined company jumped by 10,000, and the effort to economize on space began anew.

The other trend that kept messenger employment steady during the 1920s and 1930s was the increased use of messengers for accessory consumer services—especially as door-to-door sales agents and temporary office laborers. Though such efforts originated as ways to capture the work of "idle" messengers in a time-varied telegraph working day, the direct income and customer contacts from such work helped carry the telegraph system through decades of increasing competition and uncertainty. Contrary to the standard tale of universal telephone service and swift airmail service gradually demolishing the telegraph messenger service, messengers in these years were used in new ways, especially in nationwide advertising campaigns that had them delivering everything from breakfast cereal to Italian soap. Certainly these schemes should be seen as telegraph industry innovations as well.

But without the burden of young messengers, Western Union was incredibly adept at trimming its workforce over the next quarter century, shedding over forty thousand workers—four-fifths of its postwar employees (see figure 10.2). And in 1971, WU finally succeeded in its Depression-era goal of purchasing AT&T's lucrative, messenger-replacing Timed Wire Exchange (TWX) system—41,000 teleprinters that sat in the offices of customers—for about $89 million (though integrating TWX with WU's own Telex would cost another $50 million). The traditional telegraph balance of labor versus machinery was slowly overturned, and a slimmer WU became more of the "mechanism" that company president Norvin Green had desired one hundred years before.[16]

But did the modernization and job-trimming succeed in the long run? As much as WU tried to keep up with new information technology, it seemed to take two steps backward with every step forward. In 1974, the company launched the first domestic U.S. satellite communications system, Westar; however, that same year, *Business Week* argued that "much of the TWX-Telex system may already be obsolete." Even the company's balance sheet was contradictory: "To shareholders, Western Union reported a 1973 net income of $28.1 million," but "For federal tax purposes . . . it reported a $19.7 million deficit," which *Business Week* argued more accurately reflected the company's fortunes.[17]

One stark example of the fading WU fortunes was that the company actually tried to advertise as a new service a practice that fifty years before would have been considered a failure of the telegraph system. In 1970, WU began "Mailgram" service in twelve cities at $1.60 for one hundred words; by 1974, the service was available around the United States at $2 for 100 words (still cheaper than a full-rate one-hundred-word telegram, which by then cost nearly $18). The Mailgram idea was a partnership between WU and the newly reorganized U.S. Postal Service. A customer filed a Mailgram with WU, either by FAX tieline, in person, or by telephone. The message was then transmitted by

Figure 10.2. *U.S. telegraph employees, 1929–1975*

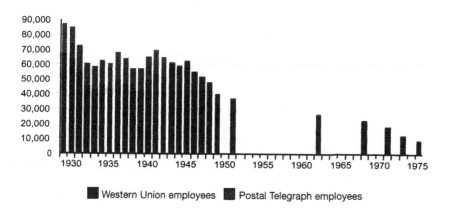

Western Union employees ■ Postal Telegraph employees

Sources: Craypo (1979), 299; WU, CPB statement (1950).
Note: Blank years indicate missing data.

WU to the post office nearest the recipient, where the postal service delivered the message as first-class mail. No time limits were in place once the post office received the message, so no refunds could be demanded. This move firmly divided WU's services into telephoned telegrams and mailed telegrams, absolving WU of the need to ever physically deliver a message or to employ messengers. Western Union proudly advertised that Mailgrams were "improving the nation's communications." But as detailed in chapter 7, telegrams had been delivered by mail for over half a century, with the practice considered a last resort when messenger delivery failed.[18]

Mailgram service was the final step in making the messengers obsolete, as new WU president Russell McFall said in 1971: "folklore notwithstanding, uniformed messenger service in our business is an anachronism. It is neither as rapid as telephone delivery of messages, nor as economical as the combination of telephone and subsequent printed copy delivery by mail. It is—as the nation becomes increasingly suburbanized—expensively difficult to provide." Given the form of the new mailed and phoned "telegrams," some questioned whether the telegraph industry provided any kind of service at all. In a 1974 exposé of WU entitled *Western Union: The Reluctant Messenger*, Congressman Benjamin S. Rosenthal declared, "What was once a rapid written communications system has—in the past 25 years—deteriorated into an 'indirect telephone service' which is much slower than the direct service offered by the telephone companies and, in some cases, slower even than the post office." Once again, besides speed and price, service coverage was an issue. Rosenthal believed that at most, only half of all U.S. residences could even receive messenger delivery, based on the number of WU offices and cooperating agencies that existed. Employment figures supported Rosenthal's claim as well: in 1972, out of WU's 15,937 nationwide employees, only 729 were motor messengers (4 percent of their workforce) and only 226 were foot or bicycle messengers (1 percent).[19]

Western Union tried to respond with more innovative programs, again in partnership with the postal service. In 1978, WU and the postal service agreed to jointly provide what they called "Electronic Computer Originated Mail" (ECOM). This was really a bulk mailing service rather than a point-to-point e-mail system, since messages were capped at two pages, and had to be submitted in batches of two hundred or more. But still, it was a new service. The message would be entered into a computer by the client and submitted via leased data lines to WU; next, WU would transmit the message to twenty-five participating post offices; finally, the offices would print the messages and deliver them via normal first-class mail. Prices ranged from 30¢ to 55¢ per message. Unfortunately for WU, when the post office applied for Postal Rate Commission approval of the new service, the FCC and the Justice Department worried about antitrust issues. The question hinged on whether ECOM was a new communications service under FCC jurisdiction, or whether it was a logical extension of postal service, under sole jurisdiction of the post office. In the end, ECOM never got off the ground.[20]

Western Union spent the 1980s, a decade of personal computers, office FAX machines, and local area networks, searching for investors and struggling to stay afloat. In 1981, aerospace manufacturing multinational Curtiss-Wright purchased nearly 22 percent of WU. But three years later, after losing $42 million on the investment, Curtiss-Wright sold its stake. That same year, WU stopped paying stock dividends. In 1987, Brooke Ltd., the holding company for, among other businesses, cigarette-maker Liggett, purchased 53 percent of WU. At this time telegrams accounted for only 10 percent of WU's revenue, but Brooke hoped WU's teletypewriter services, mailgrams, and money-order services could still be profitable. Severing the connection to the telegraph even further, WU changed its name that year from Western Union Telegraph Company to Western Union Corporation, gaining a new chairman and a new CEO charged with radically modernizing the company's business (again).[21]

They did not succeed. As the 1990s began, WU faced bankruptcy, even after selling its "EasyLink" E-mail service, its Telex service, and its other packet-switched services to AT&T for $180 million. Western Union shareholders voted to change the name of the company once again, this time to New Valley Corporation, both to hide the fact that the once great Western Union was going bankrupt, and to protect the brand image of the sole Western Union product that was still making money, the money-transfer service of the Western Union Financial Services subsidiary. Over the next few years, New Valley underwent chapter eleven bankruptcy reorganization, and sold Western Union

Financial Services to First Data Corporation, where it still survives. New Valley emerged from bankruptcy as a different firm, "transformed into the parent company of an investment banking and brokerage business and the owner of a business engaged in the ownership and management of commercial real estate." Among other holdings, New Valley bought 28 percent of Brazilian airplane maker Empresa Brasileira de Aeronautica S.A. In this way, Western Union, the company that transcended space with electricity, had come full circle, from communications back to transportation.[22]

Today, one can still send a telegram, but the service is the same in name only. The message costs $9.95 (payable by credit card), is delivered anywhere in the United States (although "certain post office boxes and rural routes may not be deliverable"), arrives the following day (if received before 6:00 P.M. Eastern Time, excluding weekends), and is limited to one thousand characters (a single page). Sending such a telegram using the modern technology of the Internet takes only twenty or so easy and straightforward steps. You can do it too:

1. Access First Data's "Western Union" World Wide Web site at http://www.westernunion.com/.
2. Click on the "eCommunications" button.
3. Click on the "Western Union® Telegram^SM" button.
4. Read about what the telegram is, and then click "Send."
5. Read a lengthy legal "Terms and Conditions" agreement defining the service, and click "I accept."
6. Provide a name, E-mail address, home address, phone number, and password (case-sensitive).
7. Read a lengthy online privacy statement, and click "I have read Western Union's online privacy statement."
8. Choose whether or not to receive "future offers and promotions from Western Union and its affiliated companies."
9. Confirm all the above information and click "Submit."
10. Print out the randomly generated nine-digit "customer ID" that is then assigned to you (or scribble it down if you don't have a printer, because you will need it in a moment).
11. Click the button that says "Click here to sign in."
12. Reenter your nine-digit customer ID and your password.
13. Click on "eCommunications" (again).
14. Click on "Western Union® Telegram^SM" (again).
15. Click on "Send" (again).
16. Click "I accept" on the "Terms and Conditions" (again).
17. Enter the recipient name, address, and the text of the telegram you want to send (not exceeding one page).

18. Enter your credit card information for payment.
19. Verify that your order is correct and click "Submit."
20. Sit back and wait for your telegram to arrive.

Does it work? Sure enough, my telegram arrived almost exactly 24 hours from the time I sent it, driven to my door by an employee of the national overnight-delivery service Airborne Express and handed to me without any mention of its telegraphic significance. The yellow cardboard envelope bore the words "Official Telegram" and sported a rather blurry photo of a smiling messenger boy, circa 1930 (with the crisp text "Western Union" digitally superimposed on the boy's cap). Inside, my telegram was laser-printed on a single sheet of yellow pin-feed computer paper. And no, I must admit, I did not tip the middle-aged "messenger boy."[23]

What this lengthy (and admittedly not very subtle) commentary on the twenty-first-century "telegram" is meant to suggest is that part of Western Union's demise may have been tied to the passing of the messenger boy. Believing the messenger model to be an outdated vestige of the "bug" days of Morse telegraphy, by 1974 WU had closed 77 percent of all offices that had existed in 1945, such that in over half of the country, telegrams could not be delivered by hand, only phoned. That year, WU handled around 21.7 million telegrams, accounting for under 18 percent of their revenues, but they only employed 779 messengers nationwide—116 of those in New York City alone, still the company's service core.[24]

Perhaps such figures meant that telegrams were now delivered more efficiently, with neither the delays nor the costs associated with messengers—be they children or adults. But just what *was* a telegram without a messenger? It was a yellow envelope that came in the mail with every other envelope, costing a few dollars when regular postage only cost a few cents. Or it was a phone call delivered by a WU telephone operator, indistinguishable from a direct business-to-business phone call from an in-house secretary? Clearly, a telegram was no longer an affordable luxury, no longer a special event, and no longer an occasion for a brave and bold messenger to track you down in the midst of your busy day and hand you an urgent communication with a smile.

After 150 years, messenger work still persists, of a sort—evidence that today's computerized information internetwork is dependent on boundary workers just as the earlier electromechanical internetwork was. Certainly in many developing areas of the globe, where so-called Plain Old Telephone Service has not yet become universal, populations must still rely on "public call office" telephones, with foot messengers notifying patrons when phones are free, when calls have connected, and when their time in line to talk has come. But even businesses in affluent American cities still rely on swift and dependable bike messengers to shuttle materials from building to building. Regardless of the availability of voice mail, E-mail, express mail, and FAX, there are still some contracts, mock-ups, special orders, or secret formulas that just can't be conveyed by electronic means. As one *New York Times* reporter put it, "the future of the e-commerce revolution is increasingly dependent on a low-tech, often-despised grunt: the bicycle messenger."[25]

A full analysis of today's urban bike messenger labor force is beyond the scope of this chapter, and unfortunately no scholars seem to have taken up the subject. But starting in the mid-1980s, the bicoastal bike messenger worlds of New York City and San Francisco exploded into popular culture, spawning a slew of magazine articles, the Hollywood movie *Quicksilver* (in which *Footloose* star Kevin Bacon dances on his bike), the short-lived network sitcom *Double Rush*, the cyberpunk messengers of William Gibson's speculative *Virtual Light*, and the popular autobiography *The Immortal Class* by Chicago messenger and playwright Travis Culley. Though long on romance and short on facts, these sources illustrate that even though the messenger demographic may have changed, from mostly white, young teen boys to a more diverse mix of young adult men and women, bike messenger work is still a difficult yet necessary job in the informational city.[26]

The modern bike messenger service was reborn in New York City with "Can Carriers," started in 1970 to lug film canisters for the advertising industry. Fifteen years later, of the three hundred to four hundred messenger services in New York City, thirty-five relied mainly on bicyclists. By the mid-1980s, there were around four thousand bicycle messengers in New York City, and though a "top messenger" could earn $500 per week, this was hard to do at a "commission" of $3.20 per drop-off (one would have to make thirty-one drop-offs per day). A good messenger at this time might have been able to average $75 per day instead. By 1990, one reporter who actually tried to work as a bike messenger for a week in New York City figured that "A rookie can average about 15 deliveries a day and gross roughly $350 for a five-day workweek." This was for riding thirty to forty miles during a day's shift of ten to eleven hours.[27]

The parallels to the bicycle exploits of the telegraph messenger boys are striking. Though today's bikes are a far cry from the first safety bikes, many modern messengers prefer to ride high-tech but simplified track bikes, with one gear, no brakes, and no coasting, meaning that to stop they must fling their front wheels into the air, still spinning. But such bikes are expensive; a basic no-frills messenger-style bike could cost $175, with obligatory messenger bag and helmet costing extra. And for every messenger with a tricked-out track bike, there are countless others, "empty-bucket riders who run ten packages a day on beat-up Huffys." Either way, messengers must supply and maintain their own bikes, a challenge when "Some riders have lost as many as 10 bikes in 5 years to theft and damage."[28]

The working conditions are also similar, according to one messenger who described his average day in detail:

> Two cabs had tried to run me off LaSalle. I'd skitched [hitched a ride by holding on to the back bumper of] a tow truck up the Fairbanks incline. I'd dropped thirty-two packages by noon. A woman had been hit by a black Mazda that spun its back end through a wet intersection. As I'd rolled past, the young woman had not gotten up. I'd been sliding everywhere; my tires couldn't seem to stick to the road. I'd delivered a very important package to a judge; he'd closed the door behind me before he signed the manifest. I'd taken a shortcut through the down ramp of the Apparel Center parking lot, where a BMW had lurched into my path. I'd hit the brakes and been thrown like a stone from its sling, landing only a few feet from the car (thank god for helmets). I'd come within a few breaths of a fistfight with a security guard at 200 Madison; I remained calm. I'd run over some guy's toes as he tried to scurry across the State Street traffic. I kept my tempo.

Like the telegraph boys, modern messengers face dangerous urban streets, made all the more perilous through their own (often illegal) quest for speed. Yet the disdain of customers and security guards alike could be just as much a danger to a messenger's all-important "tempo."[29]

Not all messenger jobs are so hectic. A small courier company might employ only one rider to transition packages between suburban office parks and downtown skyscrapers for a small group of clients—not the "anywhere, anytime" on-demand package delivery of other messenger firms. Here a messenger might earn a guaranteed minimum of $300 per week, but the messenger's relationship to this firm might be as a "1099," a term coined from the tax form used by "independent contractor" employees. A 1099 receives no benefits and has no taxes withheld. To find a more secure employment contract, one must graduate from the world of the 1099 independent contractor to the more selective (in terms of experience and references) world of the "W-4" employee, eligible for tax withholding and worker's compensation. But in W-4 courier jobs, there is often no guaranteed weekly minimum, and all deliveries are paid on commission. Here a messenger might earn about $3 on each delivery, or about $100 per day with a delivery average of over thirty packages per shift.[30]

In the early 1990s, in a scene reminiscent of the 1880s ADT boys, many New York City bike messengers, facing low wages, harsh conditions, and few benefits, were considering whether to unionize. One member of the Independent Couriers Association in New York City lamented that "there is simply too much apathy and transiency to keep a group going by being based on tradtional shop rep, committees or unions," and instead reported, "while we've gone against the companies for not providing workers' compensation insurance, we've also hit the city for regulating us and police harassment, plus we've linked up with groups promoting bicycling and alternative transportation energy schemes." Still, by 1994 New York City messengers made as little as $2.50 per package, with average pay hovering around $300 to 400 per week for eleven-hour days. The riders received no health benefits, sick leave, or vacation days, even though, according to one, each messenger averaged one accident a year.[31]

Today's messengers face other century-old issues in a new context as well. At the end of the twentieth century, once again the new technologies and emergent commodities of a wider information internetwork—this time the Internet and its World Wide Web—have had a profound effect on the construction of messenger work. With the late-1990s "dot-com" boom of Web-based business ventures, many analysts argued that "E-business" would improve the situation of the bike messengers. A vice president at Earlybird messengers in Manhattan reported in late 1999, "we're up 15 percent thanks to e-commerce . . . we've been bombarded by dot-com companies that want to crack the New York market." Fully one-third of their messenger force, one hundred riders a day, were dedicated to delivering the seven hundred or so book orders from online retailer Barnesandnoble.com. One messenger company, Kozmo.com, even built it's own E-business, renting videos and selling snacks, CDs, and other "impulse items" online, delivering them to customers "free within the hour." With the upsurge in business, Kozmo and Earlybird were among some three hundred messenger companies in New York City by late 1999, employing around ten thousand foot, bike, and van messengers. And in an all-too-familiar attempt to stand out from the competition, some services, such as Urbanfetch.com, even put their couriers in uniforms.[32]

But when the dot-com bubble burst at the turn of the millennium, it took many of these so called desktop-to-doorstep messenger companies with it. In April 2001, the largest survivor, Kozmo.com, shut down its web site and laid off its eleven hundred workers in nine U.S. cities (a little less than half of them in Manhattan). In its three years of existence, it had managed to sign up 400,000 web-based consumers, to forge deals with web bookseller Amazon.com and coffee giant Starbucks, and to spend hundreds of millions in venture capital from eager Silicon Alley incubators. But even though it was able to turn a slim profit in December 2000 in three of its urban markets, after the Nasdaq stock index dropped the bottom out of the market for initial public offerings of risky high-tech firms, the company's investors finally pulled the plug.[33]

Yet the World Wide Web truly is a worldwide (though quite globally uneven) phenomenon, and E-commerce messengers have appeared in other areas besides New York City's Silicon Alley. In Sao Paolo, Brazil, some 200,000 motorcycle couriers, known as "motoboys," are crucial to the city's economy, with the three thousand or so courier companies that employ them generating some $100 million per year in revenue alone. In a situation similar to that of Kozmo.com, the president of Motoforte Transportes de Malotes instructed a small group of his motoboys in special techniques for delivering to new e-commerce clients who in early 2001 generated some 15 percent to 25 percent of the company's revenues: "The only person he meets face to face who represents the company is the person who delivers the package. How my professional acts will affect whether that customer orders from the site again."[34]

In still other ways do today's messengers fill the cracks of the "virtual" data network. In 1985, one author described a messenger dispatching center as "a modern, computerized office" where "Dispatchers relay the information by radio to the riders, who carry walkie-talkies in holsters." With the new portable communications technologies marketed since then, today's urban messengers use both cellular alphanumeric pagers and either public-band walkie-talkies or cell phones—with equipment and airtime rented from the courier company. They ride on bicycles designed with the latest CAD/CAM systems and built under just-in-time production systems, shipped to America on state-of-the-art container ships from the Pacific Rim. Some even carry Global Positioning System transceivers while they are tracked by sophisticated Geographical Information Systems software. And according to the "cyberpunk" speculations of author William Gibson, this is only a prelude to the biotech- and nanotech-equipped messengers of tomorrow.[35]

No matter what technology they end up using, I suspect messenger work will remain difficult and—speaking as a commuter cyclist who has been personally hit by automobiles in city traffic several times—dangerous. Nevertheless, it is for some reason comforting to me that at the close of the twentieth century, just as at its opening, humble bicycle messengers still ply the borderlands between the latest information networks, knitting worlds together with every spin of their wheels. Long live the messenger.

NOTES

ACKNOWLEDGMENTS

1. For telegraph enthusiast web sites, see: Antique Wireless Association (http://www. antiquewireless.org/), KA2MGE Telegraph Museum (http://www.netsync.net/users/gm/index. htm), Morse Telegraph Club (http://members.tripod.com/morse_telegraph_club/), Retired Western Union Employees Association (http://ourworld.compuserve.com/homepages/ haroldmills/), Telegraph Office (http://fohnix.metronet.com/~nmcewen/tel_off-page.html), and Telegraph Lore (http://www.faradic.net/~gsraven/index.shtml).

2. Greg Downey, "Information networks and urban spaces: The case of the telegraph messenger boy," *Antenna* 12:1 (1999); Greg Downey, "Running somewhere between men and women: Gender in the construction of the telegraph messenger boy," *Knowledge and Society* 12 (2000): 129-152; Greg Downey, "Virtual webs, physical technologies, and hidden workers: The spaces of labor in information internetworks," *Technology and Culture* 42:2 (2001): 209–235.

CHAPTER ONE

1. George W. Gray, "These boys are always out but never down," *American Magazine* (September 1924): 24.

2. *Telegraph Age* [*T Age*] (June 1, 1923): 270.

3. E. C. Brower, memorabilia and news clippings on his retirement (1938), Western Union Archive, Archives Center, National Museum of American History, Smithsonian Institution [WUA], series 7, box 35, folder 3.

4. E. C. Brower, memorabilia (1938); *Brooklyn Eagle* (May 1938); *New York Herald Tribune* (May 29, 1938); E. C. Brower, scrapbooks (1897–1902), WUA, series 7, box 35, folder 4; telegram from B.R. Allen to E. C. Brower (May 31, 1938), WUA, series 7, box 35, folder 3.

5. Letter from A. Simion, WU metropolitan division general manager, to E. C. Brower (1938), WUA, series 7, box 35, folder 3; E. C. Brower, scrapbooks (1897–1902).

6. Zane Grey, *Western Union* (New York: Harper and Brothers, 1939); Tom Standage, *The Victorian Internet: The remarkable story of the telegraph and the nineteenth century's on-line pioneers* (New York: Walker and Company, 1998). There is no satisfactory secondary history of telegraphy, or even of industry leader Western Union, that covers the entire period of this study. For partial histories of the telegraph in the United States, see: Alexander Jones, *Historical sketch of the electric telegraph: Including its rise and progress in the United States* (New York: G. P. Putnam, 1852); George B. Prescott, *History, theory, and practice of the electric telegraph* (Boston: Ticknor and Fields, 1860); William J. Johnston, *Telegraphic tales and telegraphic history* (New York: W.J. Johnston, 1880); James

D. Reid, *The telegraph in America, and Morse memorial* (New York: John Polhemus, 1886); Alvin Harlow, *Old wires and new waves: The history of the telegraph, telephone, and wireless* (New York: D. Appleton-Century, 1936); Robert L. Thompson, *Wiring a continent: The history of the telegraph industry in the United States, 1832–1866* (Princeton, NJ: Princeton University Press, 1947); George Shiers, ed., *The electric telegraph: An historical anthology* (New York: Arno Press, 1977); George P. Oslin, *The story of telecommunications* (Macon, GA: Mercer University Press, 1992); David Hochfelder, *Taming the Lightning: American telegraphy as a revolutionary technology, 1832–1860* (Ph.D. diss., Case Western Reserve University, 1999).

7. Mickey Rooney's first appearance as a messenger boy was in the musical *I Like It That Way* (USA, 1934) at age fourteen; his most famous messenger appearance was as fourteen-year-old Homer in the 1943 film version of William Saroyan's wartime novel *The Human Comedy* (New York: Harcourt, Brace, 1943) at age twenty-three. William "Billy" Benedict, best known for his role of Skinny in the East Side Kids films (and then Whitey as the gang transformed into the Bowery Boys), became the quintessential messenger boy in American film, appearing in twelve such roles between his first appearance in *Captain January* (USA, 1936) at age nineteen and his last in *A Boy, a Girl, and a Dog* (USA, 1946) at age twenty-nine. Benedict reprised the role (uncredited) in the film *Funny Girl* (USA, 1968) when he was fifty-one years old. On Western Union product placement in film, see: George P. Oslin, *One man's century: From the Deep South to the top of the Big Apple* (Macon, GA: Mercer University Press, 1998), 68.

8. American District Telegraph Company [ADT], Philadelphia, minute books and annual reports (1878–1907) [Philadelphia records], WUA, 1996 addendum, box 4, folder 2.

9. U.S. Bureau of the Census, *Decennial Census of the United States* [US Census] (Washington, D.C.: U.S. GPO, 1870–1950). Estimates from 1870 to 1900 are based on an 8 to 9 percent portion of the broader category of "messenger and errand and office boys" (based on the 1910 count). These are conservative estimates at best; in 1884, Western Union president Norvin Green reported that his company alone employed and subcontracted nearly 8,000 messengers nationwide. Alba M. Edwards, "Comparative occupation statistics for the United States, 1870 to 1940," in *Sixteenth Census of the United States: 1940*, "Population," U.S. Bureau of the Census (Washington, D.C.: U.S. GPO, 1943); letter from Norvin Green to Nathaniel P. Hill (March 17, 1884), WUA, 1993 addendum, series C, box 51, folder 2. For other telegraph census data see: U.S. Bureau of the Census, *Telephones and telegraphs* and *Census of electrical industries* (Washington, D.C.: U.S. GPO, 1902–1932).

10. Western Union stopped using messenger boy images in its full-page print advertisements in *Telegraph and Telephone Age* around 1950; messengers disappeared from the WU *Annual Report* around the same time.

11. On the history of U.S. office work, see: William H. Leffingwell, *Scientific office management* (Chicago: A. W. Shaw, 1917); C. Wright Mills, *White collar: The American middle classes* (New York: Oxford University Press, 1951); Alfred D. Chandler, Jr., *The visible hand: The managerial revolution in American business* (Cambridge, MA: Belknap Press, 1977); Jürgen Kocka, *White collar workers in America, 1890–1940: A social-political history in international perspective*, trans. Maura Kealey (London: Sage Publications, 1980); Margery Davies, *Women's place is at the typewriter: Office work and office workers, 1870–1930* (Philadelphia: Temple University Press, 1982); Stuart Blumin, *The emergence of the middle class: Social experience in the American city, 1760–1900* (New York: Cambridge University Press, 1989); Lisa Fine, *Souls of the skyscraper: Female clerical workers in Chicago, 1870–1930* (Philadelphia: Temple University Press, 1990); Oliver Zunz, *Making America corporate: 1870–1920* (Chicago: University of Chicago Press, 1990); Sharon H. Strom, *Beyond the typewriter: Gender, class, and the origins of modern American office work, 1900–1930* (Urbana: University of Illinois Press, 1992).

12. Besides gathering firsthand messenger accounts and reminiscences from the primary literature of telegraph journals and books, about two dozen former messenger boys who worked from the 1920s to the 1950s contributed their recollections to this book. While these accounts are by no means representative (especially since they originate from the select group of messengers who were

able to forge long-term careers within Western Union), they at least allow one chorus of the voice of the messengers to be heard in the historical record. On the risks and rewards of qualitative interview methods, see: Erica Schoenberger, "The corporate interview as a research method in economic geography," *Professional Geographer* 43:2 (1991): 180–89; Neil Sutherland, "When you listen to the winds of childhood, how much can you believe?" *Curriculum Inquiry* 22:3 (1992): 235–56; and the journal *Oral History Review*.

13. "History of technology," as used here, refers not just to the history of artifacts, but the relationship of artifacts to social context. See: John M. Staudenmaier, *Technology's storytellers: Reweaving the human fabric* (Cambridge, MA: MIT Press, 1985); Langdon Winner, *The whale and the reactor: A search for limits in an age of high technology* (Chicago: University of Chicago Press, 1986); Joel A. Tarr and Gabriel Dupuy, eds., *Technology and the rise of the networked city in Europe and America* (Philadelphia: Temple University Press, 1988); George Basalla, *The evolution of technology* (New York: Cambridge University Press, 1989); Stephen H. Cutcliffe and Robert C. Post, eds., *In context: History and the history of technology* (Bethlehem, PA: Lehigh University Press, 1989); Thomas P. Hughes, *American genesis: A century of invention and technological enthusiasm, 1870–1970* (New York: Viking, 1989); Stuart W. Leslie, *The cold war and American science: The military-industrial-academic complex at MIT and Stanford* (New York: Columbia University Press, 1993); Merritt Roe Smith and Leo Marx, eds., *Does technology drive history? The dilemma of technological determinism* (Cambridge, MA: MIT Press, 1994); Carroll W. Pursell, *The machine in America: A social history of technology* (Baltimore: Johns Hopkins University Press, 1995); Chris H. Gray, ed., *Technohistory: Using the history of American technology in interdisciplinary research* (Malabar, FL: Krieger, 1996); and the journals *History and Technology* and *Technology and Culture*.

"Human geography," as used here, refers to approaches that consider the dimensions of spatiality and temporality to be fundamental to all political-economic social processes of production, consumption, and reproduction (as opposed to other subfields of geography such as human ecology, cultural geography, or humanist geography). See: Derek Gregory and John Urry, eds., *Social relations and spatial structures* (London: Macmillan, 1985); Richard Peet and Nigel Thrift, eds., *New models in geography: The political-economy perspective*, 2 vols. (London: Unwin Hyman, 1989); Derek Gregory, *Geographical imaginations* (Cambridge, MA: Blackwell, 1994); John Agnew, David N. Livingstone, and Alisdair Rogers, eds., *Human geography: An essential anthology* (Cambridge, MA: Blackwell, 1996); and the journals *Antipode*, *Environment and Planning D: Society and Space*, and *Progress in Human Geography*.

14. On the social construction of technological systems, see: Thomas P. Hughes, *Networks of power: Electrification in Western society, 1880–1930* (Baltimore: Johns Hopkins University Press, 1983); Thomas P. Hughes, "The evolution of large technological systems," in *The social construction of technological systems: New directions in the sociology and history of technology*, ed. Wiebe Bijker, Thomas P. Hughes, and Trevor Pinch (Cambridge, MA: MIT Press, 1987); Ruth Schwartz Cowan, "The consumption junction: A proposal for research strategies in the sociology of technology," in Biker, Hughes, and Pinch, eds. (1987), 261–80; David Nye, "Shaping communications networks: Telegraph, telephone, computer," *Social Research* 64 (1997): 1067–91. On the debate over concepts of "social construction" and "social shaping" of science and technology, see: Peter L. Berger and Thomas Luckmann, *The social construction of reality: A treatise in the sociology of knowledge* (Garden City, NY: Doubleday, 1966); Donald MacKenzie and Judy Wajcman, eds., *The social shaping of technology* (Philadelphia: Open University Press, 1999); Bruno Latour, *Science in action* (Cambridge, MA: Harvard University Press, 1987).

15. On the history of information technologies, see: Richard R. John, "American historians and the concept of the communications revolution," in *Information acumen: The understanding and use of knowledge in modern business* ed. Lisa Bud-Frierman(London: Routledge, 1994), 98–110; Janet Abbate, *Inventing the Internet* (Cambridge, MA: MIT Press, 1999); Paul E. Ceruzzi, *A history of modern computing* (Cambridge, MA: MIT Press, 2000); Alfred D. Chandler, Jr., and James W. Cortada, eds., *A nation transformed by information: How information has shaped the United States from colonial times to the present* (New York: Oxford University Press, 2000).

16. On the social production of space and time, see: David Harvey, "Between space and time: Reflections on the geographical imagination," *Annals of the Association of American Geographers* 80:3 (1990): 418–34; Henri Lefebvre, *The production of space*, trans. Donald Nicholson-Smith (Cambridge, MA: Blackwell, 1991); Scott Kirsch, "The incredible shrinking world? Technology and the production of space," *Environment and Planning D: Society and Space* 13 (1995): 529–55.

17. On the geography of information technologies, see: Ronald F. Abler, *The geography of intercommunications systems: The postal and telephone systems in the United States* (Ph.D. diss., University of Minnesota, 1968); Aharon Kellerman, *Telecommunications and geography* (New York: Halsted Press, 1993); Stephen Graham and Simon Marvin, *Telecommunications and the city: Electronic spaces, urban places* (New York: Routledge, 1996); Peter J. Hugill, *Global communications since 1844: Geopolitics and technology* (Baltimore: Johns Hopkins University Press, 1999); James O. Wheeler, Yuko Aoyama, and Barney Warf, eds., *Cities in the telecommunications age: The fracturing of geographies* (New York: Routledge, 2000).

18. On operator labor in the telegraph industry, see: Vidkunn Ulriksson, *The telegraphers: Their craft and their unions* (Washington, D.C.: Public Affairs Press, 1953); Charles Craypo, "The impact of changing corporate structure and technology on telegraph labor, 1870–1978," *Labor Studies Journal* 3 (1979): 283–307; Edwin Gabler, *The American telegrapher: A social history, 1860–1900* (New Brunswick, N.J.: Rutgers University Press, 1988); Melodie Andrews, "'What the girls can do': The debate over the employment of women in the early American telegraph industry," *Essays in Economic and Business History: Selected Papers from the Economic and Business Historical Society* 8 (1990): 109–20; Paul Israel, *From machine shop to industrial laboratory: Telegraphy and the changing context of American invention, 1830–1920* (Baltimore: Johns Hopkins University Press, 1992).

19. On the history of service work in the United States, see: Susan Porter Benson, *Counter cultures: Saleswomen, managers, and customers in American department stores, 1890–1940* (Urbana: University of Illinois Press, 1986); Robin Leidner, *Fast food, fast talk: Service work and the routinization of everyday life* (Berkeley: University of California Press, 1993). On the study of labor and management practices in production work, see: Michael Burawoy, "The anthropology of industrial work," *Annual Review of Anthropology* 8 (1979): 231–66; Sandra Wallman, ed., *Social anthropology of work* (New York: Academic Press, 1979); Aihwa Ong, "Gender and labor politics of postmodernity," *Annual Review of Anthropology* 20 (1991): 279–309; Randy Hodson, "The worker as active subject: Enlivening the 'New sociology of work,'" in *The new modern times: Factors reshaping the world of work* ed. David B. Bills (Albany: State University of New York Press, 1995), 253–80; Stephen R. Barley and Julian E. Orr, eds., *Between craft and science: Technical work in U.S. settings* (Ithaca, NY: IRL Press, 1997). The best work in this field recognizes that the idea of "work" itself is a socially constructed practice, not a problem-free universal economic category; work produces values and social relations, not just commodities.

20. Here "uneven geography" means that various differences between local sites of a national network—consumer populations, labor markets, environmental conditions, and urban infrastructure—inevitably affect how that network functions as a whole. For more on uneven geographical development, see: Neil Smith, *Uneven development: Nature, capital, and the production of space* (New York: Blackwell, 1984); Erica Schoenberger, *The cultural crisis of the firm* (Cambridge, MA: Blackwell, 1997).

21. On theories of space and time as they relate to power and discipline, see: Michel Foucault, *Discipline and punish: The birth of the prison*, trans. from the French by Alan Sheridan (New York: Vintage, 1995).

22. This interpretation relies on a Marxian analysis that posits that the basic social relation of capitalist production is an unequal power relation between the two roles of owner/manager and waged worker: the owner/manager is able to decide where, when, and how to invest the social surplus (though this may be subject to state regulation); the worker is constructed as a commodity to be purchased for a given wage (at the extreme, ignoring the worker's myriad life needs beyond a subsistence wage); the labor process is subject to increasing control and rationalization (with the danger of deskilling and/or disempowering the worker); and the members of the labor market are

forced to compete against each other for jobs (tending to drive down worker wages). For more on Marxian theories of commodification and labor, see: Karl Marx, *Capital*, vol. 1, trans. Ben Fowkes (New York: Penguin Books, 1976 [1867]); E. P. Thompson, *The making of the English working class* (London: Penguin Books, 1991 [1963]); Harry Braverman, *Labor and monopoly capital: The degradation of work in the twentieth century* (New York: Monthly Review Press, 1974); David Harvey, *The limits to capital* (Cambridge, MA: Blackwell, 1982).

23. In the terminology used here, "gender" is a social category, dependent on historically contingent notions of what behavior is considered "masculine" and what behavior is considered "feminine" by particular social groups, while "sex" is a biological category. For more on gender analysis with respect to technology, geography, and labor, see: Jo Little, Linda Peake, and Pat Richardson, *Women in cities: Gender and the urban environment* (New York: New York University Press, 1988); Beth Moore Milroy and Caroline Andrew, "Gender-specific approaches to theory and method," in *Life spaces: Gender, household, employment* ed. Caroline Andrew and Beth Moore Milroy (Vancouver: University of British Columbia, 1988); Joan Rothschild, "From sex to gender in the history of technology," in *context*, ed. Cutcliffe and Post; Ava Baron, "Gender and labor history: Learning from the past, looking to the future," in *Work engendered: Toward a new history of American labor* ed. Ava Baron (Ithaca, NY: Cornell University Press, 1991), 1–37; Nina Lerman, Arwen Mohun, and Ruth Oldenziel, "Versatile tools: Gender analysis and the history of technology," *Technology and Culture* 38 (1997): 1–10.

Similarly, the use of "maturity" is also meant to indicate a social category, as opposed to the biological category of "age." Maturity is defined by ideas of what behavior is thought appropriate for an "adult" versus what behavior is considered appropirate for a "child." Around the turn of the nineteenth century, such ideas were further complicated by the introduction of a new category of "adolescence" into the mix. For more on what is referred to here as a "maturity" analysis, see: Harvey J. Graff, ed., *Growing up in America* (Detroit: Wayne State University Press, 1987); Joseph M. Hawes, *Children between the wars: American childhood, 1920–1940* (New York: Twayne Publishers, 1997); David I. Macleod, *The age of the child: Children in America, 1890–1920* (New York: Twayne, 1998); and Elliott West and Paula Petrik, eds., *Small worlds: Children and adolescents in America, 1850–1950* (Lawrence: University of Kansas Press, 1992).

24. For more on this "internetwork" argument, see: Paul Edwards, "Y2K: Millennial reflections on computers as infrastructure," *History and Technology* 15 (1998): 7–29; Greg Downey, "Virtual webs, physical technologies, and hidden workers: The spaces of labor in information internetworks," *Technology and Culture* 42:2 (2001): 209–35.

25. On the history of U.S. child-labor and education reform, see: David Tyack, *The one best system: A history of American urban education* (Cambridge, MA: Harvard University Press, 1974); David Nasaw, *Schooled to order: A social history of public schooling in the United States* (New York: Oxford University Press, 1979); Stephanie Coontz, *The social origins of private life: A history of American families, 1600–1900* (New York: Verso, 1988); Joel Perlmann, *Ethnic differences: Schooling and social structure among the Irish, Italians, Jews, and Blacks in an American City, 1880–1935* (New York: Cambridge University Press, 1988); Ileen A. DeVault, *Sons and daughters of labor: Class and clerical work in turn-of-the-century Pittsburgh* (Ithaca, NY: Cornell University Press, 1990).

26. For more on the study of workers within the history of technological systems, see: David Brody, "The old labor history and the new: In search of the American working class," *Labor History* 20 (Winter 1979): 111–26; Philip Scranton, "None-too-porous boundaries: Labor history and the history of technology," *Technology and Culture* 29:4 (1988): 722–43.

27. On "modern" American society and its urban geographies, see: Marshall Berman, *All that is solid melts into air: The experience of modernity* (New York: Penguin Books, 1982); Stephen Kern, *The culture of time and space: 1880–1918* (Cambridge, MA: Harvard University Press, 1983); David Ward and Oliver Zunz, eds., *The Landscape of modernity: New York City, 1900–1940* (Baltimore: Johns Hopkins University Press, 1992).

28. The term *creative destruction* comes from Joseph Schumpeter, *Capitalism, socialism, and democracy* (New York: Harper and Bros., 1947).

29. For more on dialectics, see: David Harvey, *Justice, nature, and the geography of difference* (Cambridge, MA: Blackwell, 1996).

30. On so-called postmodern American society and its human environments, see: David Harvey, *The condition of postmodernity: An enquiry into the origins of cultural change* (Cambridge, MA: Blackwell, 1989); Anthony Giddens, *The consequences of modernity* (Stanford: Stanford University Press, 1990); Manuel Castells, *The rise of the network society*, vol. 1 of *The information age: Economy, society and culture* (Cambridge, MA: Blackwell, 1996).

CHAPTER TWO

1. William J. Johnston, *Lightning flashes and electric dashes: A volume of choice telegraphic literature, humor, fun, wit, and wisdom* (New York: W. J. Johnston, 1877), 104.

2. WU, *Western Union Messenger Service* [messenger manual] (New York: WU, 1910), Hagley Museum and Library pamphlet collection.

3. George B. Prescott, *History, theory, and practice of the electric telegraph* (Boston: Ticknor and Fields, 1860), 53; Gerald W. Brock, *The telecommunications industry: The dynamics of market structure* (Cambridge, MA: Harvard University Press, 1981), 55–56; William J. Johnston, *Telegraphic tales and telegraphic history: A popular account of the electric telegraph—its uses, extent, and outgrowths* (New York: W. J. Johnston, 1880), 36; JoAnne Yates, "The telegraph's effects on nineteenth century amarket and firms," *Business and Economic History* 15 (1986): 152. William F. Cooke and Charles Wheatstone in England also developed and patented an electric telegraph independently. For more on European telegraph history, see: Jeffrey Kieve, *The electric telegraph: A social and economic history* (London: David and Charles, 1973); Frank McCourt, *Angela's ashes: A memoir* (New York: Scribner, 1996).

4. Magnetic Telegraph Company [MT], *Articles of association and charter from the state of Maryland of the Magnetic Telegraph Company: Together with the office regulations and the minutes of the meetings of stockholders and board of directors* (New York: Chatterton and Crist, 1847). On the early history of the telegraph, see: James D. Reid, *The telegraph in America, and Morse memorial* (New York: John Polhemus, 1886); Alvin F. Harlow, *Old wires and new waves: The history of the telegraph, telephone, and wireless* (New York: D. Appleton-Century, 1936); Robert L. Thompson, *Wiring a continent: The history of the telegraph industry in the United States, 1832–1866* (Princeton, N.J.: Princeton University Press, 1947). Besides Magnetic's Morse network, others based on patents by Royal House and Alexander Bain were also constructed.

5. MT (1847), 16 (emphasis in original); Reid (1886), 125.

6. Alexander Jones, *Historical sketch of the electric telegraph: Including its rise and progress in the United States* (New York: George P. Putnam, 1852), 87; *Shaffner's Telegraph Companion* (May 1854); *New York Herald* (November 20, 1851), Henry O'Reilly papers, series II, box 31, New-York Historical Society; Allan Pred, *Urban growth and city-systems in the United States, 1840–1860* (Cambridge, MA: Harvard University Press, 1980), 11, 144–46, 155. For example, in 1840, New York, Philadelphia, Boston, and Baltimore together held 5 percent of the U.S. population, but accounted for 23 percent of all post office receipts.

7. Jeffrey L. Covell, "New Valley Corporation," in *International directory of company histories*, vol. 17 (Chicago: St. James, 1988–95), 345; James M. Herring and Gerald C. Cross, *Telecommunications: Economics and regulation* (New York: McGraw-Hill, 1936), 2; Thompson (1947), 289, 373; Harlow (1936), 332; Reid (1886), 534–35; Maury Klein, *The life and legend of Jay Gould* (Baltimore: Johns Hopkins University Press, 1986), 196.

8. Reid (1886), 534–35; WU, Executive Committee Meeting Minutes (1864–1948) [executive minutes], Western Union Archive, Archives Center, National Museum of American History, Smithsonian Institution [WUA], 1993 addendum, series B, boxes 8–30, book A, 15; WU, *Corporate history of the Western Union Telegraph Company* (New York: WU, 1914), 37; Thompson (1947), 421–25; *New York Times* [*NYT*] (April 5, 1931).

9. Harlow (1936), 250. For more on uneven geographical development, see: Neil Smith, *Uneven development: Nature, capital, and the production of space* (New York: Blackwell, 1984); Erica Schoenberger, *The cultural crisis of the firm* (Cambridge, MA: Blackwell, 1997).

10. Thompson (1947), 213. The telegraph actually preceded the railroad in many places, but after 1850 the two were usually intertwined, as the telegraph provided both a way for the railroad to more safely handle its trains using only one track (saving money on capital investments and repairs) and a way to coordinate the loading and unloading of cargo, especially to and from river barges, so that costly boxcar space could be maximized. Pred (1980); Thompson (1947), 207–12.

11. Thompson (1947), 213.

12. *Journal of the Telegraph* [*JoT*] (September 20, 1882), 178; John P. Abernethy, *The modern service of commercial and railway telegraphy* (St. Louis: J. P. Abernethy, 1894), 244; U.S. Senate, "Investigation of Western Union and postal telegraph-cable companies," Senate document 725, 60th Congress, 2nd session (Washington, D.C.: U.S. GPO, 1909), 20.

13. In the early twentieth century, there was also an annual spatial and temporal rhythm to telegraph service coverage: during winter months, when vacationers moved to Florida and California, and shipments of early vegetables and produce from these same areas to the northern cities were coordinated, the companies set up temporary offices in southern towns; during summer months, when vacationers traveled to northern mountains and seashores, and many northern rural areas brought their harvests to market, WU closed many southern offices and opened temporary northern ones. Herring and Cross (1936).

14. Thompson (1947), 68.

15. WU, *Rules* [rulebook] (New York: WU, 1884), WUA, series 10, box 53, folder 1, 8; *ACA News* (April 1941), 6.

16. Thompson (1947), 144; Jones (1852), 87; Reid (1886), 142; Pred (1980), 272.

17. U.S. Senate (1909), 24–25; Richard B. DuBoff, "The telegraph in nineteenth-century America: Technology and monopoly," *Comparative Studies in Society and History* 26 (1984): 580–81.

18. Richard B. DuBoff, "Business demand and the development of the telegraph in the United States, 1844–1860," *Business History Review* 54 (1980): 466–67; Gardiner G. Hubbard, "Government control of the telegraph," *North American Review* (December 1883).

19. Wayne E. Fuller, *The American mail: Enlarger of the common life* (Chicago: University of Chicago Press, 1972), 43; Thompson (1947), 42–47; Johnston (1880), 40.

20. Jones (1852), 110–15; DuBoff (1984), 574; Klein (1986), 68; Joel A. Tarr with Thomas Finholt and David Goodman, "The city and the telegraph: Urban telecommunications in the pre-telephone era," *Journal of Urban History* 14:1 (1987): 42.

21. George P. Oslin, *The story of telecommunications* (Macon, GA: Mercer University Press, 1992), 340; Thompson (1947), 252; Edward Highton, *The electric telegraph: Its history and progress* (London: John Weale, 1852); Jones (1852), 105; Frederick Leland Rhodes, *Beginnings of telephony* (New York: Harper and Bros., 1929), 147–49; U.S. Senate (1909), 24–25; Edwin Gabler, *The American telegrapher: A social history, 1860–1900* (New Brunswick, NJ: Rutgers University Press, 1988), 189.

22. Jones (1852), 102.

23. DuBoff (1984); Thompson (1947), 369–70; U.S. Senate (1909), 12; WU, "Annual report of the president of the Western Union Telegraph Company to the stockholders" [annual report] (New York: WU, 1869), 38.

24. Paul Israel, *From machine shop to industrial laboratory: Telegraphy and the changing context of American invention, 1830–1920* (Baltimore: Johns Hopkins University Press, 1986), 80; Charles Barnard, "The telegraph of to-day," *Harper's New Monthly Magazine* (October 1881), 707.

25. Prescott (1860), 93. In 1866 WU rules stated "*Operators not perfectly competent to do so,* are warned against attempting to receive messages by sound." WU, *Rules, Regulations, and Instructions* [rulebook] (Cleveland: Sanford and Hayward, 1866), 27.

26. Abernethy (1894), 132.

27. Israel (1986), 66; *Telegrapher* (September 1, 1866); WU, record books of George Prescott (1867–1912) [Prescott records], New York Public Library manuscript collection.

28. Reid (1886), 539, 544; WU annual report (1869); Israel (1986), 80; *JoT* (November 1, 1870), 275.

29. Melodie Andrews, "'What the girls can do': The debate over the employment of women in the early American telegraph industry," *Essays in Economic and Business History: Selected papers from the Economic and Business Historical Society* 8 (1990): 109–20.

30. Reid (1886), 273–74.

31. WU, rulebook (1866), 37–38.

32. Ralph E. Berry, *The work of juniors in the telegraph service*, Part-Time Education series no. 10, bulletin no. 7 (Berkeley: University of California Division of Vocational Education, 1922), 155.

33. WU, *Rules and instructions for the information and guidance of the employes of the Western Union Telegraph Company* [rulebook] (New York: WU, 1870), WUA, series 2, box 9, folder 13; WU, messenger manual (1910); Berry (1922), 25–37, 155, 163.

34. ADT, *Directory* (New York: ADT, 1875, 1884, 1891); *Operator and Electrical World* [*O&E World*] (March 3, 1883), 134.

35. WU, rulebook (1866), 38; WU, rulebook (1870).

36. Stephen H. Norwood, *Labor's flaming youth: Telephone operators and worker militancy, 1878–1923* (Urbana: University of Illinois Press, 1990), 27. On load-leveling in technological systems, see: Thomas P. Hughes, "The evolution of large technological systems," in *The social construction of technological systems: New directions in the sociology and history of technology*, ed. Wiebe Bijker, Thomas P. Hughes, and Trevor Pinch (Cambridge, MA: MIT Press, 1987).

37. Harlow (1936), 244; Andrew Carnegie, *Autobiography of Andrew Carnegie* (Boston: Houghton-Mifflin, 1920), 43–55; WU, Prescott records.

38. Reid (1886), 120; *JoT* (June 1, 1868), 2.

39. Reid (1886), 742.

40. WU, Harrisburg, Pennslyvania, office ledgers (1864–1902) [Harrisburg records], WUA, series 3, box 11, folders 2–3.

41. This count of the payroll numbers leaves out operators hired on a temporary basis and operator hours earned through overtime, so the operator wage figures are a conservative estimate.

42. Reid (1886), 133.

43. Reid (1886), 231–32; Jones (1852).

44. Harlow (1936), 243–44.

45. *JoT* (September 15, 1870), 240; *JoT* (May 15, 1870); *JoT* (January 15, 1872); *JoT* (August 15, 1878), 246.

46. Harlow (1936).

47. *JoT* (August 15, 1873), 280; *Telegrapher* (October 24, 1874), 253.

48. *JoT* (March 16, 1881), 90; *JoT* (December 16, 1881), 371; *JoT* (January 1, 1882), 5. Similar arguments were used to differentiate the work (and skills) ascribed to male operators versus female operators.

49. Prescott (1860), 90, 347; Abernethy (1894), 30–34.

50. Abernethy (1894), 10; W. H. Jackson, personal interview (1998); *Telegrapher* (January 28, 1871), 177.

51. *JoT* (June 15, 1868), 3 (emphasis in original); *The Pennsylvania and New Jersey Telegraph Instruction Company* (1884), Hagley Museum and Library pamphlet collection.

52. Gabler (1988), 63, 176; Israel (1986).

53. U.S. Bureau of the Census, *Decennial Census of the United States* [U.S. Census] (Washington, D.C.: U.S. GPO, 1870–1950).

CHAPTER THREE

1. American District Telegraph Company [ADT], *Directory* (New York: ADT, 1875).

2. William J. Johnston, *Telegraphic tales and telegraphic history: A popular account of the electric telegraph—its uses, extent, and outgrowths* (New York: W. J. Johnston, 1880), 32; James D. Reid, *The telegraph in America, and Morse memorial* (New York: John Polhemus, 1886), 815.

3. Paul Israel, *From machine shop to industrial laboratory: Telegraphy and the changing context of American invention, 1830–1920* (Baltimore: Johns Hopkins University Press, 1986), 100–6.

4. Israel (1986), 109; ADT, *American District Telegraph Co. [Directory]* (San Francisco: ADT, 1883); William Maver, Jr., *American telegraphy and encyclopedia of the telegraph* (New York: Maver Publishing, 1909), 371.

5. "The telegraph," *Harper's New Monthly Magazine* 47 (1873), 332–60, reprinted in George Shiers, ed., *The electric telegraph: An historical anthology* (New York: Arno Press, 1977); Edward A. Calahan, "The district telegraph," *Electrical World and Engineer* (March 16, 1901), 438–39; Joel A. Tarr with Thomas Finholt and David Goodman, "The city and the telegraph: Urban telecommunications in the pre-telephone era," *Journal of Urban History* 14:1 (1987): 49. ADT call-boxes could also be used to summon police and fire departments, and are the forerunner of today's home security systems.

6. Israel (1986), 122.

7. Western Union Telegraph Company [WU], *Corporate history of the Western Union Telegraph Company* (New York: 1914), 197; *Telegrapher* (March 15, 1873), 66; *Telegrapher* (September 15, 1873), 227.

8. WU, Executive Committee Meeting Minutes (1864–1948) [executive minutes], Western Union Archive, Archives Center, National Museum of American History, Smithsonian Institution [WUA], 1993 addendum, series B, boxes 8–30, book B, 301, 306, 323, 351, 353.

9. WU, executive minutes, book B, 355.

10. *Telegrapher* (April 25, 1874), 100–1; ADT, *Tariff Directory* (New York: ADT, 1884), 156–57; *Telegrapher* (October 9, 1875), 241–42; Reid (1886), 804–5; *Telegrapher* (April 12, 1873), 90; *Telegrapher* (July 12, 1873), 173; *Telegrapher* (October 16, 1875), 25; *Telegrapher* (December 4, 1875), 293; *New York Times [NYT]* (October 13, 1875); *NYT* (August 10, 1880); *NYT* (October 20, 1880); *Journal of the Telegraph [JoT]* (May 16, 1879); *JoT* (November 1, 1880), 325; Johnston (1880), 65; George P. Oslin, *The story of telecommunications* (Macon, GA: Mercer University Press, 1992), 260; *Operator* (November 1, 1881), 405.

11. ADT, Philadelphia, minute books and annual reports (1878–1907) [Philadelphia records], WUA, 1996 addendum, box 4, folder 2; G. T. Archibald, "The American telegraph systems" (typescript speech, 1929), 8.

12. Letter from Norvin Green to Nathaniel P. Hill, March 17, 1884, WUA, 1993 addendum, series C, box 51, folder 2; *Telegraph Age [T Age]* (December 16, 1902), 522; *Commercial Telegraphers' Journal [CTJ]* (March 1910), 80; *NYT* (March 9, 1910).

13. Reid (1886), 716; Bell Telephone Company of Pennsylvania, *Telephone, telegraph, and wireless systems in Philadelphia* (Philadelphia: Chamber of Commerce Educational Committee, 1917), in Hagley Museum and Library pamphlet collection; WU vice president W. S. Fowler, quoted in George W. Gray, "These boys are always out but never down," *American Magazine* 98 (September 1924), 25; ADT, *Directory* (1883).

14. William Hard, "'De kid wot works at night,'" *Everybody's Magazine* (January 1908), 26; Gray (1924), 24.

15. ADT, *Tariff Directory [Directory]* (New York: ADT, 1884).

16. *NYT* (March 4, 1883); Mack A. Moore, "The temporary help service industry: Historical development, operation, and scope," *Industrial and Labor Relations Review* 18 (1965): 556. For more on contingent office labor, see: Richard S. Belous, *The contingent economy: The growth of the temporary, part-time, and subcontracted workforce* (Washington, D.C.: National Planning Association, 1989).

17. U.S. Bureau of the Census, *Decennial Census of the United States* [US Census] (Washington, D.C.: U.S. GPO, 1870–1950). New York City also had the most female messengers and the most messengers under age fifteen, two measures discussed in chapter 6.

18. Joseph P. Kett, *Rites of passage: Adolescence in America, 1790 to the present* (New York: Basic Books, 1977), 151–52; Joel Perlmann, "After leaving school: The jobs of young people in Providence, R.I., 1880s-1915," in *Schools in cities: Consensus and conflict in American educational history*, ed. Ronald K. Goodenow and Diane Ravitch (New York: Holmes and Meier, 1983); Ileen A. DeVault, *Sons and daughters of labor: Class and clerical work in turn-of-the-century Pittsburgh* (Ithaca, NY: Cornell University Press, 1990).

19. U.S. Bureau of the Census, *Bulletin 69: Child labor in the United States* (Washington, D.C.: U.S. GPO, 1907), 163. Oddly, 97 percent of the wives in these 6,400 families were reported as being unemployed, but it is unclear if waged home work was counted.

20. For more on ways of defining and analyzing "class" in U.S. history, see: C. Wright Mills, *White collar: The American middle classes* (New York: Oxford University Press, 1951); Herbert G. Gutman, "Work, culture, and society in industrializing America, 1815–1919," *American Historical Review* 78:3 (1973): 531–88; Paul Boyer, *Urban masses and moral order in America: 1820–1920* (Cambridge, MA: Harvard University Press, 1978); David Brody, "The old labor history and the new: In search of the American working class," *Labor History* 20 (1979): 111–26; Roy Rosenzweig, *Eight hours for what we will: Workers and leisure in an industrial city, 1870–1920* (Cambridge: Cambridge University Press, 1983); Stuart Blumin, "The hypothesis of middle-class formation in nineteenth-century America: A critique and some proposals," *American Historical Review* 90:2 (1985): 299–338; Jacqueline Jones, *Dispossessed: America's underclasses from the Civil War to the present* (New York: Basic Books, 1992); Sharon H. Strom, *Beyond the typewriter: Gender, class, and the origins of modern American office work, 1900–1930* (Urbana: University of Illinois Press, 1992).

21. *JoT* (April 20, 1887); *T Age* (February 16, 1920); *WU News* (February 1916), 11.

22. *JoT* (June 2, 1873); *JoT* (November 16, 1877), 338; *WU News* (January 1915), 10; *T Age* (October 16, 1912), 662; *JoT* (December 1, 871); Abraham Burstein, *The ghetto messenger: Sixty tales of a unique seventy year old telegraph messenger "boy"* (Freeport, NY: Books for Libraries Press, 1971 [1928]); *T Age* (June 1, 1905), 217; Henry Miller, *Tropic of Capricorn* (New York: Grove Press, 1961), 18.

23. *WU News* (January 1916), 10; *T Age* (November 16, 1903), 563; B. Josanne, personal correspondence (1999).

24. U.S. Census (1910–50); U.S. Bureau of the Census, *The statistical history of the United States from colonial times to the present* (Stamford, CT: Fairfield Publishers, 1965).

25. *Telegrapher* (October 9, 1875), 241–42; *JoT* (February 16, 1879), 55.

26. ADT, Philadelphia records.

27. *Telegrapher* (December 4, 1875), 293; ADT, Philadelphia records.

28. WU, executive minutes, book A, 466.

29. WU, executive minutes, book C, 217; WU, executive minutes, book C, 303–4. Importantly, creating these tighter links with the district companies not only helped WU compete in the A&P battle, but provided WU with a way to experiment with a new technology, the telephone (covered in chapter 8).

30. Maury Klein, *The life and legend of Jay Gould* (Baltimore: Johns Hopkins University Press, 1986), 277; WU (1914), 197; Oslin (1992); WU, executive minutes, book E, 241–42, 253–54, 328, and book C, 560, 574, 600; Israel (1986), 122–23.

31. *NYT* (April 10, 1881); *Operator* (June 1, 1881), 205.

32. Klein (1986), 280–81; Oslin (1992), 199; J. Warren Stehman, *The financial history of the American Telephone and Telegraph Company* (Boston: Houghton Mifflin, 1925), 147–49.

33. Klein (1986), 310–11; Oslin (1992), 236.

34. Vidkunn Ulriksson, *The telegraphers: Their craft and their unions* (Washington, D.C.: Public Affairs Press, 1953), 142; Oslin (1992); H.H. Harrison, "Telegraphy in America," *Telegraph and Telephone Journal* (April 1922), 108.

35. *Postal Telegraph Magazine* [*PT*] (November 1925), 9.

36. Roy B. White, *Telegrams in 1889—and since!* (Princeton, NJ: Princeton University Press, 1939); Oslin (1992).

37. *NYT* (March 5, 1901); White (1939); Klein (1986), 385; Israel (1986), 155; U.S. Senate, "Investigation of Western Union and Postal Telegraph-Cable Companies," Senate document 725, 60th Congress, 2nd session (Washington, D.C.: U.S. GPO, 1909), 27; Alexander James Field, "The magnetic telegraph, price and quantity data, and the new management of capital," *Journal of Economic History* 52:2 (1992): 401–2.

38. WU, executive minutes, book E, 325.

39. WU, executive minutes, book D.

40. *NYT* (February 16, 1887); *NYT* (February 20, 1887); *PT* (January 1915), 15–16.

41. Norvin Green, "Are telegraph rates too high?" *North American Review* (November 1889), 571–72.

42. Ulriksson (1953), 20–22.

43. Ulriksson (1953), 58, 68, 80, 88. Such practices would lead to a nationwide strike in 1907, discussed in chapter 9.

44. Harry Braverman, *Labor and monopoly capital: The degredation of work in the twentieth century* (New York: Monthly Review Press, 1974); Frederick W. Taylor, *The principles of scientific management* (New York: Harper and Bros., 1911); ADT, Philadelphia records.

45. WU, executive minutes, book P, 837, and book Q, 12, 74.

46. Miller (1961), 21; *Dots and Dashes* [*D&D*] (February 1934); H. W. Drake and Hobart Mason, "The development and improvement in the telegraphic art," typewritten report (New York: WU, 1931), WUA, series 1, box 1, folder 4, 32.

47. EFS correspondence (1998); W. H. Jackson personal interview (1998).

48. H. D. McTavish, personal correspondence (1998).

49. Hard (1908), 180; *Electrical World* [*E World*] (June 5, 1886), 264.

50. Gray (1924), 180; EFS correspondence (1998).

51. WU, executive minutes, book P, 326; U.S. Senate, "Investigations of conditions in the wire communications industry," hearing before a subcommittee of the Committee on Interstate Commerce, United States Senate, 75th Congress, 3rd session, on S. Res. 247 (Washington, D.C.: U.S. GPO, 1938), 10, 25.

52. "Annual Christmas address of the messengers of the Western Union Telegraph Company" (Philadelphia: Beresford Printing, 1880[?]); *E World* (January 3, 1885), 2; *NYT* (1923–30).

53. Ralph Edward Berry, *The work of juniors in the telegraph service*, Part-Time Education Series no. 10, bulletin no. 7 (Berkeley: University of California Division of Vocational Education, 1922), 25–37, 155; *NYT* (August 18, 1887); *NYT* (February 17, 1887); *NYT* (February 18, 1887); *NYT* (August 7, 1880).

54. *JoT* (September 16, 1877); *NYT* (February 17, 1887); *NYT* (February 18, 1887); *NYT* (February 20, 1887); Harry Golden, *The right time: An autobiography* (New York: Putnam, 1969), 56.

CHAPTER FOUR

1. Western Union Telegraph Company [WU], *The Western Union Messenger* [messenger manual] (New York: WU, 1946), 15.

2. Henry Miller, *Tropic of Capricorn* (New York: Grove Press, 1961), 19.

3. James D. Reid, *The telegraph in America, and Morse memorial* (New York: John Polhemus, 1886), 132; *New York Illustrated News* (August 20, 1853), from Henry O'Reilly papers, series 2, box 31, New York Historical Society manuscript collection; WU, *Rulebook* (1866), 40 (emphasis in original).

4. *Telegrapher* (October 9, 1875), 241–42; American District Telegraph Company [ADT] Philadelphia minute books and annual reports (1878–1907) [Philadelphia records], Western Union Archive, Archives Center, National Museum of American History, Smithsonian Institution [WUA], 1996 addendum, box 4, folder 2; *Electrical World* [*E World*] (November 7, 1885), 189.

5. For more on metaphor, see: George Lakoff and Mark Johnson, *Metaphors we live by* (Chicago: University of Chicago Press, 1980).

6. *Journal of the Telegraph* [*JoT*] (May 15, 1869), 134; WU, *Western Union Messenger Service* [messenger manual] (New York: WU, 1910), Hagley Museum and Library pamphlet collection. For more on mechanical metaphors applied to worker bodies, see: Anson Rabinbach, *The human motor: Energy, fatigue, and the origins of modernity* (Berkeley: University of California Press, 1990); Mark Seltzer, *Bodies and machines* (New York: Routledge, 1992).

7. "Ask your mother to use care in washing the shirts to prevent fading." Rule 4, WU Messenger Manual (1910). For more on bodily and internalized measures of discipline, see: Michel

Foucault, *Discipline and punish: The birth of the prison*, tran. Alan Sheridan (New York: Vintage, 1995).

8. M. Brady Mikusko, *Carriers in a common cause: A history of letter carriers and the NALC* (Washington, D.C.: National Association of Letter Carriers, 1989), 45; *Telegrapher* (March 5, 1870), 221; *Telegrapher* (May 6, 1871).

9. ADT St. Louis messenger record book (1874–78) [St. Louis records], WUA, 1996 addendum, box 4, folder 1.

10. *Telegrapher* (February 7, 1874), 35; *JoT* (May 15, 1874), 152; WU, Executive Committee Meeting Minutes (1864–1948) [WU Executive minutes], WUA, 1993 addendum, series B, boxes 8–30, book C, 374; *Telegrapher* (January 2, 1875), 5.

11. Mike J. Rivise, *Inside Western Union* (New York: Sterling, 1950), 125; *Telegrapher* (May 9, 1874), 112–13.

12. *Telegrapher* (October 24, 1874), 257; WU, executive minutes, book D, 194, 218, 376–79, and book P, 143. These entries may not represent total uniform purchases, and are a conservative estimate only.

13. *Operator* (March 15, 1882), 98; *E World* (December 26, 1885), 265; *Postal Telegraph Magazine* [*PT*] (January 1915), 15–16.

14. KPA correspondence (1998–99); U.S. Sentate, "Investigations of conditions in the wire communications industry," hearing before a subcommittee of the Committee on Interstate Commerce, United States Senate, 75th Congress, 3rd session, on S. Res. 247 (Washington, D.C.: U.S. GPO, 1938), 25.

15. *Telegrapher* (May 3, 1873), 114–15; *Telegrapher* (June 21, 1873), 156; *Telegrapher* (July 12, 1873), 173–74.

16. *Telegrapher* (October 9, 1875), 241–42; Reid (1886), 806; *Dots and Dashes* [*D&D*] (January 1928); *Telegraph and Telephone Age* [*T&T Age*] (March 1, 1923), 123.

17. *Telegrapher* (July 12, 1873), 173–74.

18. Priscilla Ferguson Clement, "The city and the child, 1860–1885," in *American Childhood: A research guide and historical handbook* ed. Joseph M. Hawes and N. Ray Hiner (Englewood Cliffs, NJ: Greenwood Press 1985), 242; E. P. Thompson, "Time, work-discipline, and industrial capitalism," *Past and Present* 38 (December 1967).

19. "Regulations: Atlantic and Ohio Telegraph; Pittsburgh" (December 1846), Henry O'Reilly papers, series II, box 31, New York Historical Society.

20. WU, *Western Union Messenger Service* [messenger manual] (New York: WU, c. 1910), Hagley Museum and Library pamphlet collection, back cover. For more on the physical spaces of office work, see: Linda McDowell, *Capital culture: Gender at work in the city* (Oxford: Blackwell, 1997); Oliver Zunz, *Making America corporate: 1870–1920* (Chicago: University of Chicago Press, 1990).

21. George W. Gray, "These boys are always out but never down," *American Magazine* 98 (September 1924): 24–26ff; George P. Oslin, *The story of telecommunications* (Macon, GA: Mercer University Press, 1992), 205–6; *JoT* (February 15, 1875), 49–51; *Telegrapher* (February 13, 1875), 37.

22. *JoT* (October 16, 1876), 307; *JoT* (April 16, 1877), 113; *JoT* (May 1, 1877), 135; *JoT* (February 1, 1878), 34; Charles Penrose, *Newcomb Carlton, 1869–1953, of Western Union* (New York: Newcomen Society, 1956), 14–15.

23. *T Age* (July 16, 1900), 300–1.

24. Henry Miller, *Moloch, or this gentile world* (New York: Grove Press, 1992), 17–19; Henry Miller, *Tropic of capricorn* (New York: Grove Press, 1961), 29. Miller's first novel, which no longer survives today, was actually a story of thirteen bizarre messengers entitled *Clipped Wings*, a sort of anti–Horatio Alger story of the messenger boys.

25. *Telegrapher* (April 03, 1875), 71, 83; *Operator* (February 15, 1882), 82.

26. (Mrs.) W. L. Murdoch, "Conditions of child employing industries in the South," *Child Labor Bulletin* [*CLB*] (May 1913), 124–41.

27. Joel A. Tarr with Thomas Finholt and David Goodman, "The city and the telegraph: Urban telecommunications in the pre-telephone era," *Journal of Urban History* 14:1 (1987): 1–2; Sam Bass Warner, Jr., *Streetcar Suburbs: The process of growth in Boston, 1870–1900* (Cambridge, MA: Harvard University Press, 1962).

28. ADT, *Directory* (1875, 1884); Reid (1886); Dave Mote, "ADT Security Systems, Inc.," in *International Directory of Company Histories*, vol. 12 (Chicago: St. James, 1988–95), 9.

29. Reid (1886), 684; *New York Times* (December 29, 1880); *NYT* (December 30, 1880).

30. *Operator* (February 1, 1881), 45; former messenger Jimmy Duggan quoted in *PT* (November 1925), 8–10.

31. William Hard, "'De kid wot works at night,'" *Everybody's Magazine* 18 (January 1908): 37; *NYT* (February 6, 1910); Miller (1992), 18; Robert Ferguson, *Henry Miller: A life* (New York: W. W. Norton, 1991), 63.

32. *JoT* (March 15, 1869), 1; Robert Smith, *Merry wheels and spokes of steel: A social history of the bicycle* (San Bernardino, CA: Borgo Press, 1995), 13–14, 25, 35; Jay Pridmore and Jim Hurd, *The American bicycle* (Osceola, WI: Motorbooks International, 1995), 38; David B. Perry, *Bike cult: The ultimate guide to human-powered vehicles* (New York: Four Walls Eight Windows, 1995).

33. Smith (1995), 48–49; *JoT* (December 1894).

34. *T Age* (February 16, 1897), 76; *T Age* (January 1, 1894), 34; *T Age* (February 16, 1895), 72; *T Age* (June 16, 1895), 257; *T Age* (January 1, 1896), 17.

35. *T Age* (September 01, 1895), 350.

36. Stephen Kern, *The culture of time and space, 1880–1918* (Cambridge, MA: Harvard University Press, 1983), 216; Bruce Epperson, "Failed colossus: Strategic error at the Pope Manufacturing Company, 1878–1900," *Technology and Culture* 41 (April 2000): 319; *CTJ* (July 1904), 6; *CTJ* 4 (1906), 195.

37. Letter from C. J. Fogarty, Sales Manager, Westfield Manufacturing Co., to E. C. Brower, WU, June 1, 1938, from WU, E. C. Brower scrapbooks (1897–1938) [Brower scrapbooks], WUA, series 7, box 35, folders 3–4; Pridmore and Hurd (1995), 127; K. P. Akins, personal correspondence (1999).

38. WU, executive minutes.

39. *JoT* (March 15, 1869) 1; *T Age* (January 1, 1898), 7; Philip Davis, *Street-land: Its little people and big problems* (Boston: Small, Maynard and Co., 1915), 162; *T Age* (1895), 466.

40. E. Holberton, personal correspondence (1998); *CTJ* (August 1944), 7; *Business Week* (May 20, 1939).

41. Pridmore and Hurd (1995), 107; Smith (1995), 247.

42. G. K. Welner, personal correspondence (1998).

43. *JoT* (October 1, 1877), 291.

44. Viviana Zelizer, *Pricing the priceless child: The changing social value of children* (New York: Basic Books, 1985), 32, 35, 37, 40; Davis (1915), 34.

45. *Telegrapher* (May 20, 1871) 306; Miller (1992), 19; Ralph E. Berry, *The work of juniors in the telegraph service*, Part-Time Education Series no. 10, bulletin no. 7 (Berkeley: University of California Division of Vocational Education, 1922); *PT* (September 1924), 25.

46. Zelizer (1985), 35; *PT* (May 1921) 18; Gray (1924), 24.

47. E. Falborn, correspondence (1998–99).

48. K. P. Akins, personal correspondence (1999).

49. WU, executive minutes, book P, 246, 331, 568; H. L. Carraway, personal interview (1998).

50. Ellen Nathalie Matthews, "Accidents to telegraph messengers," *Monthly Labor Review* 38 (1934): 14–31.

51. *NYT* (December 6, 1934); *T World* (April 1934), 20; *T World* (May 1938), 5; *T World* (August 1936).

52. U.S. Senate (1938), 9.

53. WU, "Statement of T. B. Gittings, Assistant Vice President before the Senate Committee on Education and Labor on S. 1349." (1945), 9–10.

CHAPTER FIVE

1. *Telegraph Age* [*T Age*] (April 16, 1902), 166–69.

2. Richard R. John, "Recasting the information infrastructure for the industrial age," in *A nation transformed by information: How information has shaped the United States from colonial times to the present* ed. Alfred D. Chandler Jr., and James W. Cortada (New York: Oxford University Press, 2000), 81–82.

3. Robert L. Thompson, *Wiring a continent: The history of the telegraph industry in the United States, 1832–1866* (Princeton, NJ: Princeton University Press, 1947), 221, 242; Richard B. DuBoff, "Business demand and the development of the telegraph in the United States, 1844–1860," *Business History Review* 54 (Winter 1980): 468; Alexander Jones, *Historical sketch of the electric telegraph: Including its rise and progress in the United States* (New York: George P. Putnam, 1852), 91.

4. George P. Oslin, *The story of telecommunications* (Macon, GA: Mercer University Press, 1992), 192.

5. Tom Standage, *The Victorian Internet: The remarkable story of the telegraph and the nineteenth century's on-line pioneers* (New York: Walker and Company, 1998), 177; Roy B. White, *Telegrams in 1889—and Since!* (Princeton, NJ: Princeton University Press, 1939), 17.

6. Oslin (1992), 192–93; U.S. Senate, *Investigation of Western Union and Postal Telegraph-Cable companies*, Senate document 725, 60th Congress, 2nd session (Washington, D.C.: U.S. GPO, 1909), 10–11.

7. Paul Israel, *From machine shop to industrial laboratory: Telegraphy and the changing context of American invention, 1830–1920* (Baltimore: Johns Hopkins University Press, 1986), 100; Oslin (1992), 202; Richard B. DuBoff, "The telegraph in nineteenth-century America: Technology and monopoly," *Comparative Studies in Society and History* 26 (1984): 578.

8. U.S. Senate (1909), 22; Western Union Telegraph Company [WU], statistical notebooks prepared for Robert C. Clowry (1893–1908) [Clowry notebooks], Western Union Archive, Archives Center, National Museum of American History, Smithsonian Institution [WUA], 1993 addendum, series G, box 81, folder 4.

9. M. D. Fagen, ed., *A history of engineering and science in the Bell System*, vol. 1 (New York: Bell Telephone Laboratories, 1975), 743–46, 752; Oslin (1992), 301; William H. Leffingwell, ed., *The office appliance manual* (Chicago: National Association of Office Appliance Manufacturers, 1926), 544, 569; *Telegraph and Telephone Age* [*T&T Age*] (August 16, 1930), 377. The technical requirements for the page printer were that it be able to operate accurately over a ten-mile range at twenty words per minute or greater, using standard-width paper. This technology was different from WU's own "automatic printer" systems, which instead of printing directly onto telegram blanks printed on continuous spools of gummed tape, which an operator then ripped off and pasted onto a telegram blank. (WU's tape allowed transmission errors to be corrected more easily.)

10. Clinton A. Reed and V. James Morgan, *Introduction to business* (Boston: Allyn and Bacon, 1932), 68; E. F. Sanger, personal correspondence (1998).

11. WU, Clowry notebooks (1908); Walter P. Marshall, "A review of Western Union today as summarized before the FCC commissioners on March 9, 1961," WUA, 1993 addendum, series v.

12. Frank C. McClelland, *Office training and standards* (New York: A. W. Shaw Co., 1919); Mary F. Cahill with Agnes C. Ruggeri, *Office practice* (New York: Macmillan, 1917), 164, 176.

13. WU, Executive Committee Meeting Minutes (1864–1948) [executive minutes], WUA, 1993 addendum, series B, boxes 8–30, book A, 318; *Business Week* [*BW*] (September 16, 1931), 26–27; *Commercial Telegraphers' Journal* [*CTJ*] (March 1910), 71.

14. *BW* (September 16, 1931), 26–27; *CTJ* (March 1911), 71; Mike J. Rivise, *Inside Western Union* (New York: Sterling, 1950), 85.

15. Chester McKay, "Relationship of A.T.&T. Co. to telegraph business" (New York: WU, 1931[?]); U.S. Senate, "Study of the telegraph industry," hearings before a subcommittee of the Committee on Interstate Commerce, U.S. Senate, 77th congress, 1st session, pursuant to S. Res. 95 (Washington, D.C.: U.S. GPO, 1941).

16. *Telegrapher* (April 25, 1874), 100–1; William J. Johnston, *Telegraphic tales and telegraphic*

history: *A popular account of the electric telegraph—its uses, extent, and outgrowths* (New York: W. J. Johnston, 1880), 66; *Telegrapher* (October 9, 1875), 241–42; ADT, *Directory* (1875).

17. ADT, *Directory* (1884); ADT, *Directory* (1891), 122.

18. *Telegrapher* (November 6, 1875), 269; *Telegrapher* (October 9, 1875), 241–42; ADT, *Directory* (1884), 94; Henry Morrow Hyde, *One forty-two: The reformed messenger boy* (Chicago: Herbert S. Stone and Company, 1901).

19. Leffingwell, ed. (1926), 439; *Dots and Dashes* [*D&D*] (February 1933); WU, scrapbooks of equipment engineer H. W. Drake (1924–38) [Drake scrapbooks], WUA, series 7, box 36, folder 3 through box 37, folder 4.

20. *BW* (April 3, 1937), 20–22; *D&D* (January-February 1938); Rivise (1950), 230.

21. Frederick J. Allen, *Advertising as a vocation* (New York: Macmillan, 1919).

22. WU marketing instructions (typescript, n.d.), WUA, series 11, box 108, folder 2; strike-through indicates penciled cross-outs; underlining indicates penciled entries and changes.

23. U.S. Senate (1941), 227–28.

24. *Telegraph World* [*T World*] (May 1920), 155.

25. Henry Miller, *Moloch, or this gentile world* (New York: Grove Press, 1992), 18; *PT* (1924), 25; *Western Union News* [*WU News*] (March 1917), 167.

26. *D&D* (November 1932).

27. George W. Gray, "These boys are always out but never down," *American Magazine* (September 1924), 24.

28. Gray (1924), 26, 180–81.

29. E. A. Nicol, "Vocational guidance for Western Union messengers," *Vocational Guidance Magazine* (January 1932), 174.

30. *T Age* (August 16, 1894), 310; U.S. Senate, "Investigations of conditions in the wire communications industry," hearing before a subcommittee of the Committee on Interstate Commerce, 75th Congress, 3rd session, on S. Res. 247 (Washington, D.C.: U.S. GPO, 1938), 9.

31. James D. Reid, *The telegraph in America, and Morse memorial* (New York: John Polhemus, 1886), 136; JH interview (1999); L. J. Feucht, personal correspondence (1998).

32. *D&D* (March 1930); *T World* (March 1930), 18; *T World* (May 1930), 22; *T World* (October 1930), 20.

33. Robin Leidner, *Fast food, fast talk: Service work and the routinization of everyday life* (Berkeley: University of California Press, 1993); *T&T Age* (November 1, 1934), ii.

34. Jeff W. Hayes, *Autographs and memoirs of the telegraph* (Adrian, MI: S. F. Finch, 1916), 66; *JoT* (December 1, 1878), 358; K. P. Akins, personal correspondence (1999).

35. Ernest D. Chase, *The greeting card industry* (Boston: Belman Publishing, 1946), 10.

36. *D&D* (May 1929); White (1939), 15; Rivise (1950), 178; *BW* (September 16, 1931), 26–27.

37. *D&D* (May 1930); *D&D* (November 1937); *D&D* (June 1930); *D&D* (November-December 1938); *D&D* (December 1925).

38. W. H. Jackson, personal interview (1998); *D&D* (June 1929).

39. *D&D* (September 1927); *D&D* (September 1932).

40. Oslin (1992), 332; George P. Oslin, *One man's century: From the deep South to the top of the Big Apple* (Macon, GA: Mercer University Press, 1998), 69.

41. Rivise (1950), 166–67; *New York Times* [*NYT*] (January 1, 1940); E. Holberton, personal correspondence (1999).

42. *Telegrapher* (March 1, 1873), 59; *CTJ* (April 1904), 6; Gray (1924), 25.

43. Rivise (1950), 15; *NYT* (May 1, 1937).

44. Annteresa Lubrano, *The telegraph: A case study in the sociology of technology innovation* (Ph.D. diss., City University of New York, 1995); *BW* (September 16, 1931), 26–27.

45. Ralph E. Berry, *The work of juniors in the telegraph service*, Part-Time Education Series no. 10, bulletin no. 7 (Berkeley: University of California Division of Vocational Education, 1922), 144.

46. *D&D* (January 1932); *D&D* (February 1932).

47. Frederick Lewis Allen, *Only yesterday: An informal history of the 1920s* (New York: Harper and Row, 1964 [1931]), 181.

48. *D&D* (February 1932); *D&D* (November 1932).

49. *T&T Age* (October 1939), ii.

50. WU, advertising flyers (c. 1920), Hagley Museum and Library pamphlet collection.

51. J. Hollansworth, personal interview (1999). In 1935, new WU president Roy B. White actually proposed using plain white paper for telegraph blanks instead of the traditional yellow, in order to cut costs during the Depression, but reconsidered after "public outcry." Oslin (1998), 76.

52. *ACA News* (November 23, 1940), 5; *BW* (April 3, 1937), 20–22.

53. U.S. Senate, *To authorize a complete study of the telegraph industry*, hearings before a subcommittee of the Committee on Interstate Commerce, U.S. Senate, 76th congress, 1st session, on S. Res. 95 (Washington, D.C.: U.S. GPO, 1939), 53; *NYT* (August 24, 1945).

54. *NYT* (August 24, 1945); Rivise (1950), 167.

CHAPTER SIX

1. Horatio Alger, Jr., *Adventures of a telegraph boy, or "Number 91"* (New York: Frank F. Lovell, 1899).

2. James D. Reid, *The telegraph in America, and Morse memorial* (New York: John Polhemus, 1886), 805; *New York Times [NYT]* (March 4, 1883).

3. Reid (1886), 642–43; *Commercial Telegraphers' Journal [CTJ]* (May 1915); U.S. Bureau of the Census, *Decennial Census of the United States* [U.S. Census] (Washington, D.C.: U.S. GPO, 1870–1950). Although they were rarely employed as urban telegraph messengers, girls did perform similar kinds of errand work for other firms, especially in the clothing industry, where they delivered prepared clothes to customers and obtained raw materials from fabric stores. Girls doing such work in Boston in 1915 earned from $3 to $5 per week, working 8 to 10 hours per day. Harriet Dodge, *Survey of occupations open to a girl of fourteen to sixteen years of age* (Boston: Girls' Trade Education League, 1915).

4. U.S. Census (1910–50).

5. *Telegraph and Telephone Age [T&T Age]* (September 1942), 17; *T&T Age* (August 1943), 16; American Communications Association [ACA], brief submitted by the ACA before the U.S. Department of Labor, Wage and Hour and Public Contracts Division, in support of the recommendation of Industry Committee #69 to establish a 40¢ hourly minimum in the communication, utilities, and micellaneous transportation industries (January 17, 1944), Wisconsin Historical Society [WHS], collection 298, box 18, folder 5; *T&T Age* (April 1944), i; U.S. Census (1910–50).

6. *CTJ* (December 1903), 3. The only time women did make a dent in the messenger service came as telegraph messages themselves began to be received and sent via telephone, allowing a new workforce of female messengers to "deliver" telegrams all over the city without ever leaving their chairs (considered in chapter 7).

7. *NYT* (March 4, 1883).

8. William J. Johnston, *Telegraphic tales and telegraphic history: A popular account of the electric telegraph—its uses, extent, and outgrowths* (New York: W. J. Johnston, 1880), 66; Henry Morrow Hyde, *One forty-two: The reformed messenger boy* (Chicago: Herbert S. Stone, 1901); *Popular Science Monthly* (August 1937), 48–49.

9. Women weren't always the driving force behind such work. Having a young, presexual boy personally serve a woman instead of an adult man may have been the goal of men as well. In 1883, the *Times* gave an example of a drunken man calling a messenger to take him home to a wife who he feared would have been too "warm" had a hired man brought the husband home. *NYT* (March 4, 1883).

10. For more on "separate spheres" as both contemporary ideology and historical tool for analysis, see: Linda Kerber, "Separate spheres, female worlds, woman's place: The rhetoric of women's history," *Journal of American History* 75 (June 1988): 38; Nancy F. Cott, "On men's history and women's history," in *Meanings for manhood: Constructions of masculinity in Victorian America*, ed.

Mark C. Carnes and Clyde Griffen (Chicago: University of Chicago Press, 1990), 206–7; Elizabeth Wilson, "The invisible *flâneur*," *New Left Review* 191 (1992): 71.

11. Norman L. Munro, "Young Dilke, the messenger boy detective; or, on the trail of a band of Wall Street swindlers," *Old Cap. Collier library* 715 (New York: Munro's Publishing House, 1897), 7; Charles Morris, "Shadow Sam, the messenger boy; or, turning the tables," *Beadle's half dime library* 10:235 (New York: Beadle and Adams, 1882), 3; George Ade, "Handsome Cyril, or the messenger boy with the warm feet," *The strenuous lad's library* 1 (Phoenix: Bandar Log Press, 1903); Joseph M. Hawes, *Children in urban society: Juvenile delinquency in nineteenth-century America* (New York: Oxford University Press, 1971), 123. For more on dime novels, see: John G. Cawelti, *Apostles of the self-made man: Changing concepts of success in America* (Chicago: University of Chicago Press, 1965); Michael Denning, *Mechanic accents: Dime novels and working-class culture in America* (New York: Verso, 1998).

12. Luc Sante, *Low life: Lures and snares of old New York* (New York: Vintage, 1991), 179. The term *vice* is used here as a general term for a shifting range of entertainment activities taking place on the borders of legality and propriety, primarily conducted by and for men, but usually involving the labor of women. Such activities included not only prostitution, gambling, drinking, and drug use, but also penny cinema, musical theater, and burlesque. *Vice* was a pejorative term for what participants understood as work or leisure; it was often left to self-appointed reform groups to argue over what was vice and what was not.

13. Timothy J. Gilfoyle, "Policing of sexuality," in *Inventing Times Square: Commerce and culture at the crossroads of the world* ed. William R. Taylor (Baltimore: Johns Hopkins University Press, 1991), 298; Jeremy P. Felt, *Hostages of fortune: Child labor reform in New York State* (Syracuse, NY: Syracuse University Press, 1965), 155.

14. Gilfoyle (1991); Barbara Meil Hobson, *Uneasy virtue: The politics of prostitution and the American reform tradition* (New York: Basic Books, 1987); Mary E. Odem, *Delinquent daughters: Protecting and policing adolescent female sexuality in the United States, 1885–1920* (Chapel Hill: University of North Carolina Press, 1995); Research Committee of the Committee of Fourteen, *The social evil in New York City: A study of law enforcement* (New York: Andrew H. Kellogg Co., 1910).

15. Hobson (1987), 141–45; David Nasaw, *Children of the city: At work and at play* (New York: Oxford University Press, 1985), 82.

16. *Electrical World* [*E World*] (November 10, 1883), 172; John D'Emilio and Estelle Freedman, *Intimate matters: A history of sexuality in America* (New York: Harper and Row, 1988), 181; Felt (1965), 154–58; ADT, Philadelphia minute books and annual reports (1878–1907) [Philadelphia records], Western Union Archive, Archives Center, National Museum of American History, Smithsonian Institution [WUA], 1996 addendum, box 4, folder 2.

17. Joseph P. Kett, *Rites of passage: Adolescence in America, 1790 to the present* (New York: Basic Books, 1977), 89; Jacob Riis, *The children of the poor* (New York: Charles Scribner's Sons, 1892), 113; John Spargo, *The bitter cry of the children* (London: Macmillan, 1906).

18. Josephine C. Goldmark, "Street labor and juvenile delinquency," *Political Science Quarterly* 19:3 (1904): 417–38; William Hard, "'De kid wot works at night,'" *Everybody's Magazine* 18 (January 1908): 27.

19. Hard (1908).

20. Edward N. Clopper, *Child labor in city streets* (New York: Macmillan, 1912a); Owen R. Lovejoy, "Child labor and the night messenger service," *Survey* (May 21, 1910), 311–17; Lewis W. Hine, "Present conditions in the South," *Child Labor Bulletin* [*CLB*] (February 1914), 59–69; National Child Labor Committee, *Street-workers*, pamphlet no. 246 (New York: National Child Labor Committee, 1915); U.S. Senate, *Child labor in the District of Columbia*, hearing before a Subcommittee of the Committee on the District of Columbia, United States Senate, 66th Congress, 2nd session, on S. 3843 (Washington, D.C.: U.S. GPO, 1920). On social surveys, see: Martin Bulmer, Kevin Bales and Kathryn Kish Sklar, eds., *The social survey in historical perspective, 1880–1940* (New York: Cambridge University Press, 1991). On the NCLC, see: Walter Trattner,

Crusade for the children: A history of the National Child Labor Committee and child labor reform in America (Chicago: Quadrangle Books, 1970).

21. Edward N. Clopper, "The night messenger boy," Annals of the American Academy of Political Science supplement (July 1911): 103; Lovejoy (1910), 312–13.

22. Lovejoy (1910), 313; Nasaw (1985), 141.

23. Edward N. Clopper, "The proper standard for street trades regulation—ways and means to secure it," CLB (August 1912b), 114–18; Clopper (1912a), 104, 113. The term delinquency, like vice, was subject to shifting definitions. For example, at this time the Chicago Juvenile Court defined as "delinquent" those "who violate any law, who are incorrigible, who knowingly associate with vicious persons, who are growing up in idleness and crime, who knowingly frequent a disorderly gaming house." Under such a definition, most night messengers were legally delinquent. Sophonisba P. Breckinridge and Edith Abbott, The delinquent child and the home: A study of the delinquent wards of the Juvenile Court of Chicago (New York: Russell Sage Foundation, 1912), 11.

24. Telegraph Age [T Age] (March 16, 1895), 105; E. C. Brower scrapbooks (1897–1938) [Brower scrapbooks], WUA, series 7, box 35, folder 3.

25. Jane Addams, The spirit of youth and the city streets (Urbana: University of Illinois Press, 1972 [1909]), 28, 103; see also G. Stanley Hall, Adolescence (New York: D. Appleton and Company, 1904). Apparently, reformers never worried about messenger boys becoming prostitutes themselves, though they did fear this might happen to messenger girls. For the story of a homosexual sex scandal involving messenger boys in London, see H. Montgomery Hyde, The Cleveland Street Scandal (London: W. H. Allen, 1976).

26. Lovejoy (1910), 313; Clopper (1912a), 104–7; Chicago Vice Commission, The social evil in Chicago (Chicago: Gunthrop-Warren, 1911), 244.

27. Charles E. Rosenberg, "Sexuality, class, and role in ninteenth century America," in The American man, ed. Elizabeth H. Pleck and Joseph H. Pleck (Englewood Cliffs, NJ: Prentice-Hall, 1980), 234; Michael S. Kimmel, Manhood in America: A cultural history (New York: Free Press, 1996), 121, 412 n.8.

28. Pleck and Pleck, eds. (1980), 25; Kimmel (1996), 121. On the Boy Scouts, see David I. Macleod, "Act your age: Boyhood, adolescence, and the rise of the Boy Scouts of America," Journal of Social History 16 (1982): 3–20.

29. Addams (1972); Nasaw (1985), 116; Ernest Poole, "Waifs of the street," McClure's Magazine (May 1903), 40–44, reprinted in Robert H. Bremner, ed., Children and youth in America: A documentary history, vol. 2, 1866–1932, parts one through six (Cambridge, MA: Harvard University Press, 1971).

30. Jon M. Kingdale, "The 'Poor Man's Club': Social functions of the urban working-class saloon," in Pleck and Pleck, eds. (1980), 255–83.

31. Hobson (1987); NYT (May 2, 1910).

32. CTJ (April 1911), 105; T Age (June 1, 1904), 236.

33. Peter G. Buckley, "Boundaries of Respectability: Introductory essay," in Taylor, ed. (1991), 288; Gilfoyle (1991), 306; NYT (May 2, 1910); NYT (May 22, 1910); Florence Kelley, "The street trader under Illinois law," in The child in the city: A series of papers presented at the conferences held during the Chicago Child Welfare Exhibit, ed. Sophonisba P. Breckinridge (Chicago: Chicago School of Civics and Philanthropy, Department of Social Investigation, 1912), 290–301.

34. CLB (February 1915), 29.

35. Lovejoy (1910).

36. Kelley (1912); NYT (February 3, 1918).

37. U.S. Census (1900–50).

38. CTJ (September 1915); CTJ (May 1915). Note that it was not unusual for messengers employed by the same company in the same city to be working on slightly different shifts, or at slightly different wages, whether due to favoritism or to simple bureaucratic differences between branch offices.

39. Scott Nearing, "One district messenger," Independent (February 22, 1912), 412–13.

40. *WU News* (June 1915); *WU News* (August 1916); *WU News* (July 1917).

41. *T World* (April 1931), 23.

42. *WU News* (January 1918).

43. *CTJ* (December 1903); U.S. Senate, *Child labor in the District of Columbia*, hearing before a Subcommittee of the Committee on the District of Columbia, United States Senate, 66th Congress, 2nd session, on S. 3843 (Washington, D.C.: U.S. GPO, 1920), 77, 84.

44. Odem (1995), 121–22; Hobson (1987).

45. Henry Miller, *Moloch, or this gentile world* (New York: Grove Press, 1992); *Postal Telegraph Magazine* (June 1924), 39.

46. George W. Gray, "These boys are always out but never down," *American Magazine* 98 (September 1924), 26.

47. B. Josanne, personal correspondence (1999); WU, *Messenger manual* (1946), 15.

48. U.S. Senate, *Investigations of conditions in the wire communications industry*, hearing before a subcommittee of the Committee on Interstate Commerce, United States Senate, 75th Congress, 3rd session, on S. Res. 247 (Washington, D.C.: U.S. GPO, 1938); F. P. Bergman, personal correspondence (1998).

49. Robin Leidner, "Serving hamburgers and selling insurance: Gender, work, and identity in interactive service jobs," *Gender and Society* 5:2 (1991): 154–77.

CHAPTER SEVEN

1. American Telephone and Telegraph Company [AT&T], *Annual report of the directors of American Telephone and Telegraph Company to the stockholders for the year ending December 31, 1910* [annual report] (New York: AT&T, 1911), 53 (emmphasis in original).

2. Malcolm Willey and Stuart Rice, *Communication agencies and social life* (New York: McGraw-Hill, 1933), 210.

3. Edward A. Calahan, "The district telegraph," *Electrical World and Engineer* (March 16, 1901), 438–39. On information internetworks, see: Paul Edwards, "Y2K: Millennial reflections on computers as infrastructure," *History and Technology* 15 (1998): 7–29; Greg Downey, "Virtual webs, physical technologies, and hidden workers: The spaces of labor in information internetworks," *Technology and Culture* 42:2 (2001): 209–35.

4. U.S. House of Representatives, *Preliminary report on communications companies*, committee print on House res. 59 and House joint res. 572 (Washington, D.C.: U.S. GPO, 1934), xiii; Mike J. Rivise, *Inside Western Union* (New York: Sterling, 1950), 159; Chester McKay, "Relationship of A.T.&T. Co. to telegraph business" (manuscript, c. 1931), Western Union Archive, Archives Center, National Museum of American History, Smithsonian Institution [WUA], 1993 addendum, series G, box 81, folder 7, 2; U.S. Senate, *To authorize a complete study of the telegraph industry*, hearings before a subcommittee of the Committee on Interstate Commerce, US Senate, 76th congress, 1st session, on S. Res. 95 (Washington, D.C.: U.S. GPO, 1939), 38; Western Union Telegraph Company [WU], Executive Committee Meeting Minutes (1864–1948) [executive minutes], WUA, 1993 addendum, series B, boxes 8–30, book E, 85; Susan Porter Benson, *Counter cultures: Saleswomen, managers, and customers in American department stores, 1890–1940* (Urbana: University of Illinois Press, 1986), 20.

5. Robert L. Thompson, *Wiring a continent: The history of the telegraph industry in the United States, 1832–1866* (Princeton, NJ: Princeton University Press, 1947); Jeffrey Kieve, *The electric telegraph: A social and economic history* (London: David and Charles, 1973), 154, 161–62; Vidkunn Ulriksson, *The telegraphers: Their craft and their unions* (Washington, D.C.: Public Affairs Press, 1953), 8. On the government telegraph debates in the U.S., see: Gardiner G. Hubbard, "Government control of the telegraph," *North American Review* (December 1883); Frank Parsons, "The telegraph monopoly," *Arena* (May 1896); A. S. Burleson, *Government ownership of electrical means of communication*, letter from the postmaster general to the U.S. Senate, 63rd congress, 2nd session, document no. 399 (Washington, D.C.: U.S. GPO, 1914). Interestingly, by the early 1910s, because telegraph signals could be superimposed on dual-wire telephone circuits (but telephone sig-

nals could not be reliably sent over single-wire telegraph circuits), the post office wanted take over only the wire plant of the telephone companies, not the wires of the telegraphs.

6. Richard R. John, Jr., "Private mail delivery in the United States during the nineteenth century: A sketch," *Business and Economic History* 15 (1986), 138; Gerald Cullinan, *The United States Postal Service* (New York: Praeger, 1973), 85; M. Brady Mikusko, *Carriers in a common cause: A history of letter carriers and the NALC* (Washington, D.C.: National Association of Letter Carriers, 1989), 4–5; Wayne E. Fuller, *The American mail: Enlarger of the common life* (Chicago: University of Chicago Press, 1972), 71. For more on the early history of the post office, see: Richard R. John, *Spreading the news: The American postal system from Franklin to Morse* (Cambridge, MA: Harvard University Press, 1995).

7. WU, *Argument of William Orton on the Postal Telegraph Bill, delivered before the Committee on Post-Offices and Post-Roads of the Senate of the United States, January 20, 21, 22, and 23* (New York: WU, 1874), 34.

8. Interestingly, a messenger service organized in 1897 "in several Eastern cities" that offered a regular pick-up and delivery schedule using telegram boxes placed outside residences was charged with violating postal laws and was ordered to be discontinued. *Telegraph Age* [*T Age*] (April 1, 1897), 147.

9. *T Age* (October 16, 1903), 524.

10. Cullinan (1973), 128, 135; Vern K. Baxter, *Labor and politics in the U.S. Postal Service* (New York: Plenum Press, 1994), 51.

11. *Dots and Dashes* (April 1925).

12. William H. Leffingwell, ed., *The office appliance manual* (Chicago *Dots and Dashes*: National Association of Office Appliance Manufacturers, 1926), 451; [*D&D*] (January 1927). Of course, the businessman in this example would still have to "dictate" a telegram to someone, just as he dictated a letter, so the WU cost argument made little sense.

13. U.S. Senate, *Study of the telegraph industry*, report of the Committee on Interstate Commerce, U.S. Senate, 77th Congress, 1st session, on S.Res.95 (U.S. Committee on Interstate Commerce, 1941a), 17; U.S. Senate (1939), 9, 36, 50; U.S. Senate, *Study of the telegraph industry*, hearings before a subcommittee of the Committee on Interstate Commerce, U.S. Senate, 77th congress, 1st session, pursuant to S. Res. 95 (Washington, D.C.: U.S. GPO, 1941b), 8.

14. Baxter (1994), 52; U.S. Senate (1941a), 23.

15. WU, *Rulebook* (1866), 33; John P. Abernethy, *The modern service of commercial and railway telegraphy* (St. Louis, MO: J. P. Abernethy, 1894), 201.

16. Ulriksson (1953), 40; WU, *Rulebook* (1870).

17. Fuller (1972), 74–77; Cullinan (1973), 102–5; Willey and Rice (1933), 105–6; WU, *The Western Union and the War Labor Board: The company's position* (New York: WU, 1918), 14.

18. WU, *Rulebook* (1870); *Postal Telegraph Magazine* [*PT*] (September 1921), 15; *PT* (September 1925), 25; B. Josanne, personal correspondence (1999).

19. Cullinan (1973), 189; Fuller (1972), 73.

20. Ralph Frank, ADT history pamphlet (preproduction version, 1999), 4; *WU News* (November 1916), 93; A. M. Hartley, personal correspondence (1998); *T Age* (May 16, 1895), 193; Frank C. McClelland, *Office training and standards* (New York: A. W. Shaw, 1919), 18.

21. Mikusko (1989), 19; *American Child* (November 1921), 209; U.S. Bureau of Labor Statistics, "Technological changes and employment in the United States Postal Service," *Bulletin of the U.S. Bureau of Labor Statistics* 574 (December 1932): 58.

22. U.S. Senate, *Investigations of conditions in the wire communications industry*, hearing before a subcommittee of the Committee on Interstate Commerce, U.S. Senate, 75th Congress, 3rd session, on S. Res. 247 (Washington, D.C.: U.S. GPO, 1938), 8; Mary P. Corre, "The letter carrier," *Occupations* (April 1936), 641–43; Cullinan (1973), 153.

23. Mikusko (1989), v.

24. George D. Smith, *The anatomy of a business strategy: Bell, Western Electric, and the origins of the American telephone industry* (Baltimore: Johns Hopkins University Press, 1985), 25, 35, 38;

George P. Oslin, *The story of telecommunications* (Macon, GA: Mercer University Press, 1992), 217, 222; John Brooks, *Telephone: The first hundred years* (New York: Harper and Row, 1975).

25. Smith (1985); Joel A. Tarr with Thomas Finholt and David Goodman, "The city and the telegraph: Urban telecommunications in the pre-telephone era," *Journal of Urban History* 14:1 (1987): 51.

26. Smith (1985), 36–37; J. Warren Stehman, *The financial history of the American Telephone and Telegraph Company* (Boston: Houghton Mifflin, 1925), 14.

27. Stehman (1925), 12; Frederick L. Rhodes, *Beginnings of telephony* (New York: Harper and Bros., 1929), 50; WU, executive minutes, book D, 585; Oslin (1992), 228; Brooks (1975), 71; Smith (1985), 56, 76–77.

28. Smith (1985), 76; Abernethy (1894), 191 (emphasis in original).

29. Brooks (1975), 65; Claude S. Fischer, "'Touch someone': The telephone industry discovers sociability," *Technology and Culture* 29:1 (1988), reprinted in Marcel C. Lafollette and Jeffrey K. Stine, eds., *Technology and choice: Readings from technology and culture* (Chicago: University of Chicago Press, 1991), 93; J. Leigh Walsh, *Connecticut pioneers in telephony: The origin and growth of the telephone industry in Connecticut* (New Haven, CT: Telephone Pioneers of America, 1950), 46–48, citing the Southern New England Telephone Company archives.

30. Stephen H. Norwood, *Labor's flaming youth: Telephone operators and worker militancy, 1878–1923* (Urbana: University of Illinois Press, 1990), 27; Frederick L. Rhodes, *John J. Carty: An appreciation* (New York: privately printed, 1932), 10.

31. Walsh (1950), 74; Norwood (1990), 27; Angus Smith Hibbard, *Hello, goodbye: My story of telephone pioneering* (Chicago: A. C. McClurg and Co., 1941 [1913]), 24. Sporadic strike activity among the boys in the 1880s might also have been a reason for managers to favor girl operators; see Marion May Dilts, *The telephone in a changing world* (New York: Longmans, Green and Co., 1941), 205 n.5.

32. Norwood (1990), 28–29, 40–41; *New York Times* [*NYT*] (February 20, 1887); *T Age* (April 16, 1894), 153.

33. *T Age* (June 16, 1900), 248; William Maver, Jr., *American telegraphy and encyclopedia of the telegraph* (New York: Maver Publishing, 1909), 371.

34. Alvin Harlow, *Old wires and new waves: The history of the telegraph, telephone, and wireless* (New York: D. Appleton-Century, 1936), 386.

35. Louis Galambos, "Looking for the boundaries of technological determinism: A brief history of the U.S. telephone system," in *The development of large technical systems*, ed. Renate Mayntz and Thomas Hughes (Boulder, CO: Westview Press, 1988), 140; Alfred B. Paine, *In one man's life: Being chapters from the personal and business career of Theodore N. Vail* (New York: Harper and Bros., 1921), 238.

36. Fischer (1988), 90; Galambos (1988), 140; AT&T, *Events in telecommunications history* (Warren, NJ: AT&T Archives, 1992), 29–30.

37. Susan J. Douglas, *Inventing American broadcasting, 1899–1922* (Baltimore: Johns Hopkins University Press, 1987), 159; Paine (1921), 232–34; Brooks (1975), 134–35; *Business Week* [*BW*] (September 16, 1931), 26–27; Stehman (1925), 149; Gerald W. Brock, *The telecommunications industry: The dynamics of market structure* (Cambridge, MA: Harvard University Press, 1981), 152; T. A. Wise, "Western Union, by grace of FCC and AT&T," *Fortune* (March 1959), 116.

38. Harlow (1936), 402; AT&T (1992), 31; WU, executive Minutes, book E, 344.

39. *Commercial Telegraphers' Journal* [*CTJ*] (March 1911), 71; AT&T, *Annual Report* (1911).

40. Herbert N. Casson, *The history of the telephone* (Chicago: A. C. McClurg and Co., 1910), 282; New York State Legislature, Joint Committee to Investigate Telephone and Telegraph Companies, *Report of Joint Committee of the Senate and Assembly of the State of New York Appointed to Investigate Telephone and Telegraph Companies*, vol. 1 (Albany, NY: J. B. Lyon Co., 1910), 475–76.

41. Stehman (1925), 150.

42. Brooks (1975), 135; Stehman (1925), 140; Galambos (1988), 143; Theodore N. Vail, "Public utilities and public policy," *Atlantic Monthly* 111 (1913): 307–19.

43. Stehman (1925), 154; *PT* (January 1915), 24; *PT* (April 1915), 1; *D&D* (March 1926).

44. James M. Herring and Gerald C. Cross, *Telecommunications: Economics and regulation* (New York: McGraw-Hill, 1936), 8–9; Wise (1959), 115, 120.

45. *PT* (August 1921), 6–9. Telephone delivery was practiced in small towns and rural areas as well, but in these cases, an office clerk or a messenger might make the phone calls.

46. *PT* (August 1921), 6–9; Ralph E. Berry, *The work of juniors in the telegraph service*, Part-Time Education Series no. 10, bulletin no. 7 (Berkeley: University of California Division of Vocational Education, 1922), 15; *PT* (April 1925), 9–11; *Telegraph and Telephone Age* (November 16, 1910), 764–65.

47. *PT* (April 1925), 9–11; *PT* (August 1921), 6–9.

48. *NYT* (February 2, 1918); G. T. Archibald, "The American telegraph systems" (typescript, 1929), WUA, series 1, box 1, folder 1, 30.

49. Willey and Rice (1933), 136–38.

50. *T Age* (May 16, 1895), 193.

51. Claude S. Fischer, *America calling: A social history of the telephone to 1940* (Berkeley: University of California Press, 1992); Clinton A. Reed and V. James Morgan, *Introduction to business* (Boston: Allyn and Bacon, 1932), 56; Arthur W. Grumbine, "The Era of Morse Telegraphy," *Dots and Dashes*, the journal of the Morse Telegraphy Club (not to be confused with WU's *D&D*) (1985).

CHAPTER EIGHT

1. John I. Sowers, *The boy and his vocation* (Peoria, IL: Manual Arts Press, 1925), 99.

2. Mike J. Rivise, *Inside Western Union* (New York: Sterling, 1950), 20.

3. On labor market theory, see: Peter B. Doeringer and Michael J. Piore, *Internal labor markets and manpower analysis* (Lexington, MA: D. C. Heath, 1971); Jamie Peck, *Work-place: The social regulation of labor markets* (New York: Guilford Press, 1996). Messenger employers desired both to reap the material gains of an external labor market and to offer the psychological attraction of an internal labor market.

4. *Telegrapher* (October 9, 1875), 241–42.

5. *Journal of the Telegraph* [*JoT*] (May 15, 1870), 146; *Commercial Telegrapher's Journal* (May 1915).

6. *JoT* (June 15, 1875), 185; *Telegrapher* (09 October 1875), 241–42.

7. James D. Reid, *The telegraph in America, and Morse memorial* (New York: John Polhemus, 1886), 232, 245; *WU News* (March 1916), 1; *Telegraph Age* [*T Age*] (September 16, 1893), 335; George W. Gray, "These boys are always out but never down," *American Magazine* (September 1924), 183. For a sampling of telegrapher biographies, see John B. Taltavall, *Telegraphers of to-day: Descriptive, historical, biographical* (New York: P. F. McBreen, 1894).

8. *WU News* (May 1917), 210.

9. Robert Ferguson, *Henry Miller: A life* (New York: W. W. Norton, 1991), 63; Mary V. Dearborn, *The happiest man alive: A biography of Henry Miller* (New York: Simon and Schuster, 1991); Henry Miller, *Moloch, or this gentile world* (New York: Grove Press, 1992), 37.

10. *T Age* (May 1, 1896), 177; *WU News* (August 1917), 273; Anne S. Davis, *Occupations and industries open to children between fourteen and sixteen years of age* (Chicago: Chicago Board of Education, 1914).

11. Elizabeth F. Baker, *Technology and women's work* (New York: Columbia University Press, 1964), 67; U.S. Senate, "Investigation of Western Union and Postal Telegraph-Cable companies," Senate document 725, 60th Congress, 2nd session (Washington, D.C.: U.S. GPO, 1909), 43; WU, *Western Union Service as a Career* (New York: WU, c. 1917), Hagley Museum and Library pamphlet collection. Western Union later transfered the internal career myth to the thousands of messenger girls it hired during World War II, saying that many "are now working as telegraph sales clerks and teleprinter operators and may become branch office managers." *D&D* (August 1942); *Dots and Dashes* [*D&D*] (July 1943).

12. Reid (1886); Andrew Carnegie, *Autobiography of Andrew Carnegie* (Boston: Houghton-Mifflin, 1920), 43.

13. *Telegrapher* (May 23, 1874), 124; *Telegrapher* (May 30, 1874), 128; *Telegrapher* (October 9, 1875), 241–42. ADT had little to fear from this law, as it turned out, because it was weakly enforced. Forest C. Ensign, *Compulsory school attendance and child labor* (Iowa City, IA: Athens Press, 1921), 121.

14. *New York Times* [*NYT*] (August 8, 1880); ADT, Philadelphia minute books and annual reports (1878–1907) [Philadelphia records], WUA, 1996 addendum, box 4, folder 2.

15. For more on early-twentieth-century child-labor and education reformers, see: Walter Trattner, *Crusade for the children: A history of the National Child Labor Committee and child labor reform in America* (Chicago: Quadrangle Books, 1970); LeRoy Ashby, *Saving the waifs: Reformers and dependent children, 1890–1917* (Philadelphia: Temple University Press 1984); Margo Horn, *Before it's too late: The child guidance movement in the United States, 1922–1945* (Philadelphia: Temple University Press, 1989).

16. Joseph M. Hawes and N. Ray Hiner, eds., *American childhood: A research guide and historical handbook* (Westport, CT: Greenwood Press 1985), 621.

17. Arthur J. Jones, *The continuation school in the United States*, Bureau of Education Bulletin no. 367 (Washington, D.C.: U.S. GPO, 1907), 32.

18. Gray (1924), 181; Ernest Poole, "Waifs of the street," *McClure's Magazine* (May 1903), 40–44, reprinted in Robert H. Bremner, ed., *Children and Youth in America: A documentary history, Volume II: 1866–1932* (Cambridge, MA: Harvard University Press, 1971).

19. Owen R. Lovejoy, "Child labor and the night messenger service," *Survey* (May 21, 1910), 313–14; Edward N. Clopper, *Child labor in city streets* (New York: Macmillan, 1912), 104; Sophonisba P. Breckinridge and Edith Abbott, *The delinquent child and the home: A study of the delinquent wards of the Juvenile Court of Chicago* (New York: Russell Sage Foundation, 1912), 74. For more on the NCLC, see: Trattner (1970).

20. Raymond G. Fuller, *Child labor and the constitution* (New York: Thomas J. Crowell 1923), x; Lydia H. Crane, "The messenger boy," *Child Labor Bulletin* [*CLB*] (August 1914), 23–26.

21. Jeremy P. Felt, *Hostages of fortune: Child labor reform in New York State* (Syracuse, NY: Syracuse University Press, 1965), 52–56, 64, 71; Hawes and Hiner, eds. (1985), 622; Jones (1907), 10. For more on turn-of-the-century educational reform efforts, see: Ronald D. Cohen, "Child-saving and Progressivism, 1885–1915," in Hawes and Hiner, eds. (1985); Felt (1965); Harvey Kantor and David Tyack, eds., *Work, youth and schooling: Historical perspectives on vocationalism in American education* (Stanford, CA: Stanford University Press, 1982); Diane Ravitch and Ronald K. Goodenow, eds., *Educating an urban people: The New York City experience* (New York: Teachers College Press, 1981).

22. Jones (1907), 7–9, 82; E. O. Holland, "Child labor and vocational work in the public schools," *CLB* 1:1 (1912), 16–23.

23. Jones (1907), 91, 97–98.

24. New York City Department of Education, *Youth in school and industry: A report issued in cooperation with the continuation school principals of the city of New York* (New York: New York City Department of Education, 1935), 16–17; Franklin J. Keller, *Day schools for young workers: The organization and management of part-time and continuation schools* (New York: The Century Co., 1924), 61; Florence Kelley, "Part time schools," *CLB* 1:1 (1912), 106–12; New York City Board of Education, *The first fifty years: A brief review of progress, 1898–1948, 50th annual report of the Superintendent of Schools* (New York: New York City Board of Education, 1948), 67.

25. U.S. Bureau of Education, *Part-time education of various types*, Bureau of Education Bulletin no. 5 (Washington, D.C.: U.S. GPO, 1921), 20–22; Keller (1924), 62; New York City Department of Education (1935), 19–20; Howard G. Burdge, *Our boys* (New York: State Military Training Commission, 1921), 26.

26. Joseph F. Kett, *Rites of passage: Adolescence in America, 1790 to the present* (New York: Basic Books, 1977), 239–40; New York City Board of Education (1948), 65–66; U.S. Bureau of Education (1921), 13; J. Lynn Barnard, *Getting a living: A vocational civics reader* (Philadelphia: Franklin Publishing and Supply, 1926), 8; Keller (1924), 36.

27. Lewis A. Wilson, *Part-time school for the working youth*, Bulletin no. 756 (Albany: University of the State of New York, 1922), 11–12, 15; Lee Galloway, *Office management: Its principles and practice* (New York: The Ronald Press, 1918), 496–97.

28. Oakley Furney, *Organization and administration of part-time schools in manufacturing or mercantile establishments and in factories*, Bulletin no. 790 (Albany: University of the State of New York, 1923) 6–10.

29. Furney (1923), 8.

30. Jones (1907), 131; Arthur G. Wirth, *Education in the technological society: The vocational-liberal studies controversy in the early twentieth century* (Washington, D.C.: University Press of America, 1980), 84, 117–18; Albert J. Beatty, *Corporation schools* (Bloomington, IL: Public School Publishing Company, 1918), 44, 50–51; New York City Board of Education, *Annual Report* (1922), 1898; New York City Board of Education, *Annual Report* (1925). For more on the benefits of corporate schools to business, see: Samuel Bowles and Herbert Gintis, *Schooling in capitalist America: Educational reform and the contradictions of economic Life* (New York: Basic Books, 1976), 193–94; Harvey Kantor, "Vocationalism in American education: The economic and political context, 1880–1930," in Kantor and Tyack, eds. (1982), 14–44.

31. Carroll G. Pearse, "Child labor and the future development of the school," *CLB* (January 1912), 38–45; U.S. Commission on Industrial Relations, *Final report of the Commission on Industrial Relations* (Washington, D.C.: 1915), 271; U.S. Board of Education, *Part-time education*, 7, 16.

32. Wilson (1922), 7; William L. Ettinger, *A report of conditions in continuation schools* (New York City Board of Education, January 1924), 29, 52; Selma Berrol, "Immigrant children at school, 1880–1940: A child's eye view," in *Small worlds: Children and adolescents in America, 1850–1950*, ed. Elliott West and Paula Petrik (Lawrence: University of Kansas Press, 1992), 44, 325; William A. Bullough, *Cities and schools in the Gilded Age: The evolution of an urban institution* (Port Washington, NY: Kennikat Press, 1974), 24; Ronald D. Cohen and Raymond A. Mohl, *The paradox of progressive education: The Gary Plan and urban schooling* (Port Washington, NY: Kennikat Press, 1979), 36; Walter William Pettit, *Self-supporting students in certain New York City high schools* (New York: New York School of Social Work, 1920); New York City Department of Education, *Youth in School and Industry*, 21.

33. Keller (1924), 62; Ettinger (1924), 29; Felt (1965), 119.

34. Philip Davis, *Street-land: Its little people and big problems* (Boston: Small, Maynard and Co., 1915), 162; *Postal Telegraph Magazine* [*PT*] (February 1915), 10–11; *PT* (March 1915), 20; *PT* (June 1915), 20.

35. *PT* (November 1915), 10; *PT* (September 1915), 15–16; U.S. Senate (1909), 55.

36. *WU News* (August 1918), 473; *PT* (October 1925), 23.

37. New York City Board of Education *Journal of the Board of Education of the City of New York*, [New York City Board of Education Journal], (September 26, 1923), 2245. New York City Board of Education, Minutes of the Superintendents of Education [NYC Supts. of Ed. Minutes] (May 21–December 10, 1925; September 24, 1936).

38. NYC Supts. of Ed. Minutes, (March 16, 1928); "The Western Union Continuation School for Messenger Boys," *Elementary School Journal* 28 (June 1928): 728–29; *NYT* (June 13, 1928); E. A. Hungerford, *The trail to successful careers*, pamphlet reprinted from *Association Men*, journal of the YMCA (1926); *Telegraph and Telephone Age* [*T&T Age*] (July 1, 1929), 295.

39. Ralph E. Berry, *The work of juniors in the telegraph service*, Part-Time Education Series no. 10, Bulletin no. 7 (Berkeley: University of California Division of Vocational Education, 1922), 17; WU Executive Committee, book P, 479, 598.

40. Berry (1922), 19.

41. Berry (1922), 19; New York City Board of Education Journal, (March 10, 1926), 458.

42. WU Executive Committee, book Q, 378; "The Western Union Continuation School for Messenger Boys," *Elementary School Journal* (June 1928): 728–29; *T&T Age* (February 1, 1933), 33.

43. *D&D* (August 1928); blueprints, WUA, series 3, box 12, folder 3.

44. *WU News* (April 1915), 5; "The Western Union Continuation School for Messenger Boys," *Elementary School Journal* 28 (June 1928): 728–29.

45. WU Executive Committee; *American Child* magazine; *NYT* (March 14, 1953).; Hungerford (1926); Frederick M. Trumbull, *Guidance and education of prospective junior wage earners* (New York: J. Wiley and Sons, 1929), 25. For more on the Boy Scouts, the YMCA, and perceptions of boyhood, see: David I. Macleod, *Building character in the American boy: The Boy Scouts, Y.M.C.A., and their forerunners, 1870–1920* (Madison: University of Wisconsin Press, 1983).

46. *D&D* 8:11 (November 1932).

47. Ettinger (1924), 22, 43–45.

48. New York City Board of Education, *The first fifty years*, 103; Keller (1924), 90; *American Child* (December 1923), 4; National Industrial Conference Board, *The employment of young persons in the United States* (New York: National Industrial Conference Board, 1925), 45.

49. Jones (1907), 10; David Tyack, *The one best system: A history of American urban education* (Cambridge, MA: Harvard University Press, 1974), 177; Lawrence Cremin, *The transformation of the school: Progressivism in American education, 1876–1957* (New York: Knopf, 1961), 186–89.

50. Ettinger (1924), 100–1; Edward N. Clopper, "Why overlook the street worker?" *CLB* (May 1914), 56–58. Note: "retarded" in this context meant "children who are older than they should be for the grades they are in"; New York City elementary schools reported that 39 percent of their pupils fit this definition in 1904. Leonard P. Ayres, *Laggards in our schools: A study of retardation and elimination in city school systems* (New York: Russell Sage Foundation, 1909).

51. Keller (1924), 86, 425–26.

52. David Montgomery, *Workers' control in America: Studies in the history of work, technology, and labor struggles* (New York: Cambridge University Press, 1979), 140; Hungerford (1926).

53. Ira M. Dreese, *Personnel studies of messengers in the Western Union Telegraph Company* (Ph.D. diss., Columbia University, 1929).

54. Henry Miller, *Tropic of Capricorn* (New York: Grove Press, 1961), 21; Dreese (1929), 1, 12.

55. Gray (1924), 182; Dreese (1929), 3, 16. Yearly totals confirmed these monthly snapshots: of the 6,000 WU New York City messenger boys employed in 1928 (to fill the 2,000 or so positions) only 104 (1.7 percent) were promoted to other WU jobs.

56. Keller (1924), 85.

57. Dreese (1929), 16–19.

58. Dreese (1929).

59. E. A. Nicol, "Vocational guidance for Western Union messengers," *Vocational Guidance Magazine* (January 1932), 172.

60. Nicol (1932), 172–76; *D&D* (November 1932).

61. Grayson N. Kefauver, Victor H. Noll, and C. Elwood Drake, *Part-time secondary schools*, Bulletin 1932, no. 17, National Survey of Secondary Education, monograph no. 3 (Washington D.C.: U.S. GPO, 1933), 5.

62. Dearborn (1991), 73; Nicol (1932), 169–71; George Oslin, *Talking wires: The way of life in the telegraph industry* (Evanston, IL: Row, Peterson and Co., 1942), 4.

63. WU, *Messenger Manual* (1946), 25; E. S. Brown, personal correspondence (1998); L. J. Feucht, personal correspondence (1998); E. T. Goldsworthy, personal correspondence (1999).

64. Letter from Joseph Kehoe, ACA, to Kathleen Lenroot, Children's Bureau (August 24, 1938), WHS, collection 298, box 23, folder 4.

CHAPTER NINE

1. Western Union Telegraph Company [WU], "Statement of T.B. Gittings, Assistant Vice President before the Senate Committee on Education and Labor on S. 1349" (New York: WU, 1945).

2. *Operator and Electrical World* [*O&E World*] (April 21, 1883), 248.

3. Robert L. Thompson, *Wiring a continent: The history of the telegraph industry in the United States, 1832–1866* (Princeton, NJ: Princeton University Press, 1947), 388–91; Edwin Gabler, *The*

American telegrapher: A social history, 1860–1900 (New Brunswick, NJ: Rutgers University Press, 1988), 146; Vidkunn Ulriksson, *The telegraphers: Their craft and their unions* (Washington, D.C.: Public Affairs Press, 1953), 15–17, 19.

4. *Journal of the Telegraph* [*JoT*] (December 2, 1867), 4; George P. Oslin, *The story of telecommunications* (Macon, GA: Mercer University Press, 1992), 188, 193; Ulriksson (1953), 19.

5. Ulriksson (1953), 20, 28–29; Gabler (1988), 150.

6. James D. Reid, *The telegraph in America, and Morse memorial* (New York: John Polhemus, 1886), 547; Gabler (1988), 7.

7. Foster R. Dulles and Melvyn Dubofsky, *Labor in America: A history* (Wheeling, IL: Harlan Davidson, 1993), 122, 125–27.

8. Gabler (1988), 7, 159–61, 173; Dulles and Dubofsky (1993), 127; Ulriksson (1953), 32; *O&E World* (January 20, 1883) 38.

9. Gabler (1988), 1–26, 183; Ulriksson (1953), 49. Dulles and Dubofsky (1993), 127–39; *Electrical World* [*E World*] (July 28, 1883) 472]; *JoT* (February 20, 1887), 19.

10. *Telegrapher* (June 13, 1874), 143; *JoT* (June 15, 1874), 184; *Telegrapher* (May 29, 1875), 129, 131.

11. *New York Times* [*NYT*] (August 7–10, 1880).

12. Dulles and Dubofsky (1993), 108; *Operator* (June 1, 1881), 205; *Operator* (January 1, 1882), 3; *Operator* (December 1, 1881), 459; *NYT* (August 10–12, 1880).

13. *Operator* (April 1, 1882), 116, 123; *Operator* (July 1, 1882), 273.

14. *NYT* (March 11, 1887).

15. *NYT* (February 16–23, 1887); Dulles and Dubofsky (1993), 139.

16. *Telegraph Age* [*T Age*] (June 16, 1899), 252; *NYT* (February 6, 1910); *NYT* (July 22, 1910).

17. *Commercial Telegrapher's Journal* [*CTJ*] (December 1910), 380; H. F. J. Porter, "The strike of the messenger boys," *Survey* (December 10, 1910), 431–32; *NYT* (November 2, 1910).

18. Ulriksson (1953), 59, 61–62; Dulles and Dubofsky (1993), 149–54.

19. *CTJ* (May 1903), 2; *CTJ* (October 1903), 4; *CTJ* (December 1903), 3; *CTJ* (1906), 12; *CTJ* (1906), 88; *CTJ* (1906), 371.

20. *CTJ* (April 1905), 9.

21. *CTJ* (October 1905), 18.

22. David Montgomery, *Workers' control in America: Studies in the history of work, technology, and labor struggles* (New York: Cambridge University Press, 1979), 93; *CTJ* 5 (1907), 831, 913; Ulriksson (1953), 72–87.

23. Charles Craypo, "The impact of changing corporate structure and technology on telegraph labor, 1870–1978," *Labor Studies Journal* 3 (1979): 295; U.S. Senate, *Investigation of Western Union and Postal Telegraph-Cable companies*, Senate document 725, 60th Congress, 2nd session (Washington, D.C.: U.S. GPO, 1909).

24. Stephen H. Norwood, *Labor's flaming youth: Telephone operators and worker militancy, 1878–1923* (Urbana: University of Illinois Press, 1990), 74; Ulriksson (1953), 91; J. Warren Stehman, *The financial history of the American Telephone and Telegraph Company* (Boston: Houghton Mifflin, 1925), 151; *CTJ* (July 1914), 269–70; *NYT* (March 20, 1914); Ulriksson (1953), 104. For more on "welfare capitalism," see: Stuart D. Brandes, *American welfare capitalism, 1880–1940* (Chicago: University of Chicago Press, 1976).

25. *WU News* (July 1914), 1; Ulriksson (1953); U.S. Commission on Industrial Relations, *Final feport of the Commission on Industrial Relations* (Washington, D.C.: U.S. GPO, 1915), 19.

26. William Lazonick, *Competitive advantage on the shop floor* (Cambridge, MA: Harvard University Press, 1990), 248; Dulles and Dubofsky (1993), 217; Montgomery (1979), 96; Norwood (1990), 162.

27. Dulles and Dubofsky (1993), 218; Norwood (1990), 157; Brandes (1976), 127.

28. *NYT* (February 2, 1918); *NYT* (February 3, 1918); *NYT* (February 5, 1918).

29. *WU News* (March 1918), 390; U.S. Senate, *Study of the telegraph industry*, hearings before a subcommittee of the Committee on Interstate Commerce, U.S. Senate, 77th congress, 1st session, pursuant to S. Res. 95 (Washington, D.C.: U.S. GPO, 1941), 142–43.

30. WU, *The Western Union and the War Labor Board: The company's position* (New York: WU, 1918), 4–7.

31. WU (1918), 3, 14, 28.

32. WU (1918), 6, 33–35.

33. U.S. Senate (1941), 144.

34. Norwood (1990), 164; U.S. Senate (1941), 144–45.

35. Ulriksson (1953), 120, 125; *Telegraph World* [*T World*] (1919); Lazonick (1990), 248; Brandes (1976), 122. Also started during World War I was the American Bell Association, an AT&T company union created under the same conditions as WU's AWUE. Thomas R. Brooks, *The communications workers of America* (New York: Mason/Charter, 1977).

36. Oslin (1992), 279; Ulriksson (1953), 115, 122; Frederick Lewis Allen, *Only yesterday: An informal history of the 1920s* (New York: Harper and Row, 1964 [1931]), 47–48; Lazonick (1990), 248–50; Robert H. Zieger, *American workers, american unions* (Baltimore: Johns Hopkins University Press, 1994), 23; Dulles and Dubofsky (1993), 250; Brandes (1976); *T World* (1929).

37. Ulriksson (1953), 124; *NYT* (March 21–24, 1919).

38. *CTJ* (October 1925), 409–10, 473–77.

39. *CTJ* (October 1925), 409–10, 473–77.

40. Zieger (1994), 26; Ulriksson (1953), 133–34; *T World* (March 1940); U.S. National Labor Relations Board [NLRB], "Findings of fact in the matter of the Western Union Telegraph Company and the American Communciations Association, case no. C-344" (October 12, 1938), WHS, collection 298, box 18, folder 9.

41. *T World* (October 1932), 15; *T World* (June 1934); *T World* (March 1934), 12; *T World* (November 1934).

42. *T World* (June 1930), 1.

43. *T World* (December 1934), 10; *T World* (May 1934), 16; *T World* (June 1934).

44. American Communications Association [ACA], "Brief in support of the allegations of the complaint against the Western Union Telegraph Company before the National Labor Relations Board, case no. C-344" (August 11, 1938), WHS, collection 298, box 18, folder 9; U.S. NLRB (1938).

45. *NYT* (June 13, 1937); Ulriksson (1953), 133–34.

46. Zieger (1994), 39–40; Montgomery (1979), 164; Dulles and Dubofsky (1993), 265.

47. *NYT* (August 10, 1937); *NYT* (August 31, 1937); Mike J. Rivise, *Inside Western Union* (New York: Sterling, 1950), 223.

48. Ulriksson (1953), 139, 144–45.

49. Dulles and Dubofsky (1993), 278–86; Zieger (1994), 44.

50. Zieger (1994), 46, 51–54; Dulles and Dubofsky (1993), 288.

51. Ulriksson (1953), 139–47.

52. Ulriksson (1953), 152; *T World* (May 1940), 1.

53. ACA, "Brief on petitions for review (no. 236) and petition for enforcement (no. 237) of an order of the National Labor Relations Board in the US Circuit Court of Appeals for the Second Circuit" (April 25, 1940), WHS, collection 298, box 18, folder 9, 8.

54. *Business Week* [*BW*] (October 29, 1938), 28–29; Dulles and Dubofsky (1993), 275; Memo from D. Driesen, ACA, to M. Rathborne, J. Selly, and J. Kehoe, ACA, on "Messenger exemption from wage-hour law" (October 1939), WHS, collection 298, box 23, folder 4; letter from ACA Central Region Director Doug Ward to messengers in Detroit and Cleveland (August 23, 1938), WHS, collection 298, box 23, folder 4 (emphasis in original); U.S. Senate, *Investigations of conditions in the wire communications industry*, hearing before a subcommittee of the Committee on Interstate Commerce, United States Senate, 75th congress, 3rd session, on S. Res. 247 (Washington, D.C.: U.S. GPO, 1938), 14; *ACA News* (November 26, 1938).

55. Commercial Telegrapher's Union [CTU], "Proceedings," 18th convention (Chicago: September 11–15 1939), 16–17.

56. Ulriksson (1953), 148–50; *ACA News* (December 23, 1939).

57. *T World* (October 1939), 10–11; Ulriksson (1953), 158–60; *T World* (March 1940), 1; *T World* (May 1940), 1; U.S. NLRB, "Decision and order in the matter of the Western Union Telegraph Company and the American Communciations Association, case no. C-344" (1939), WHS, collection 298, box 18, folder 9. Again, in a parallel with the telephone industry, the National Federation of Telephone Workers was organized in 1939 to replace the former Bell company unions. Melvin J. Segal, "Industrial relations in communications: Telephone and telegraph," *Labor in postwar America*, vol. 2, ed. E. Colston Warne et al. (New York: Remsen Press, 1949), 429–48.

58. CTU (1939), 10–11.

59. CTU (1939), 11; CTU, "Proceedings," 19th convention (Toronto Ontario, October 27–31, 1941), 52.

60. *ACA News* (March 1942), 2.

61. U.S. Department of Labor, "Report and recommendations of the fact finding board appointed by order of July 11, 1946, in the dispute between the Western Union Telegraph Company and the National Coordinating Board, AFL" (August 29, 1946), 2–3; U.S. Census (1940).

62. *ACA News* (June 1941), 2, 4; *ACA News* (August 1941), 1; *ACA News* (October 1941), 1.

63. Segal (1949), 440.

64. U.S. Department of Labor (1946), 1–4; Segal (1949), 440–42.

65. Letter from ACA president Joseph P. Selly to Harry A. Millis, chair, National Labor Relations Board (August 28, 1944), WHS, collection 298, box 18, folder 11; *ACA News* (July 1944), 3; ACA, *Organizing guide for field organizers and volunteer organizers: Western Union organizing campaign* (June 26, 1944), WHS, collection 298, box 18, folder 11.

66. CTU, "Proceedings," 20th convention (Cincinnati, OH, October 18–22, 1943); *CTJ* (March 1942), 12.

67. U.S. Bureau of Labor Statistics, *Wage chronology, Western Union Telegraph Co., 1945–1953* (Washington, D.C.: U.S. GPO, 1960), 1; Segal (1949), 440–42; Ulriksson (1953), 166; *ACA News*.

68. *CTJ* (October 1945), 7; *ACA News* (April 1945), 2; WU, "Statement of T. B. Gittings," 5. Western Union did receive an exemption for employees in "telegraph agencies with telegraph revenues up to $500 per month"—small subcontracted telegraph offices. WU, "A special report to the stockholders and employees of the Western Union Telegraph Company on the future of the record communication industry" (October 20, 1949).

69. Segal (1949), 443–45. The ACA, feeling cheated out of $6 million in wages, called out seven thousand ACA members on strike in January 1946, crippling WU for five weeks. Arbitration ruled in favor of WU.

70. *ACA News* (March 1947), 1.

CHAPTER TEN

1. Russell W. McFall, *Making history by responding to its forces* (New York: Newcomen Society, 1971), 13.

2. Travis H. Culley, *The immortal class: Bike messengers and the cult of human power* (New York: Villard, 2001), xix.

3. WU Executive Committee, book P, 101; *Telegraph Age* [*T Age*] (May 1, 1925), 220; *WU Supervisory News* (July 22, 1953), 3; Thomas M. Tucker, "Tyco Laboratories, Inc.," *International directory of company histories* [*IDCH*], vol. 3 (Chicago: St. James, 1988–95), 644; T. A. Wise, "Western Union, by grace of FCC and AT&T," *Fortune* (March 1959), 216.

4. Tucker, "Tyco Laboratories," 644; Dave Mote, "ADT Security Systems, Inc.," *IDCH*, vol. 12, 9–11.

5. Wise (1959); H. H. Goldin, "Government policy and the domestic telegraph industry," *Journal of Economic History* 7:1 (1947): 53–68.

6. George P. Oslin, *The story of telecommunications* (Macon, GA: Mercer University Press, 1992), 338; *ACA News* (February 1953), 4–5; *Telegraph and Telephone Age* [*T&T Age*] (April 1947), 21; *T&T Age* (April 1949), 27; Wise (1959), 216–18. Interestingly, Marshall came not from WU management, but from WU's competitor PT, as he was PT executive vice president when his company merged with WU in 1943.

7. *T&T Age* (April 1948), 30; *Forbes* 82:11 (1958), reprinted in *WU Telegraph News* (January 1959), 8–9; Western Union exhibits 1954, Wester Union Archive, Archives Center, National Museum of American History, Smithsonian Institution, 1993 addendum, series V.

8. David Montgomery, *Workers' control in America: Studies in the history of work, technology, and labor struggles* (New York: Cambridge University Press, 1979), 166; Robert H. Zieger, *American workers, American unions* (Baltimore: Johns Hopkins University PRess, 1994), 108–11; Foxter R. Dulles and Melvyn Dubofsky, *Labor in American: A history* (Wheeling, IL: Harlan Davidson, 1993), 343–45.

9. Montgomery (1979), 169; Zieger (1994), 131–32.

10. *ACA News* (April 1950), 6; *ACA News* (October 1951), 5; *T&T Age* (August 1951), 17; American Communications Association [ACA], *The plot (. . . that failed) to destroy our union: The exposure of a Western Union scheme, documented from the McCarran Committee record* (New York: ACA, 1951); US Senate, Committee on the Judiciary, *Subversive infiltration in the telegraph industry* (Washington, D.C.: U.S. GPO, 1951), 79.

11. Dulles and Dubofsky (1993); *T&T Age* (May 1947), 26; *WU Supervisory News* (September 24, 1951); Thomas D. Durrance, "Clicking again: Western Union makes good in an electronics age," *Barron's* (May 9, 1955); WU, Statement to the president's Communications Policy Board concerning domestic record communications policies, November 17, 1950 (New York: WU, 1950) [WU CPB statement], exhibit 54–B. Between 1941 and 1946, messenger wages rose from about 30¢ per hour to 65¢ per hour; nonmessenger wages rose from 68¢ per hour to $1.16 per hour.

12. *T&T Age* (March 1949), 20; B. Bowlen, personal correspondence (1999); W. Bechtel, personal interview (1998); Federal Communications Commission, *Statistics of the communications industry in the United States*, vols. 1939–57 (Washington D.C.: U.S. GPO, 1939–57).

13. *Forbes* 82:11 (1958); Wise (1959), 218; WU, "And now . . . our next 100 years" (New York: WU, 1952), WUA, 1993 addendum, series I, box 82, folder 1; U.S. Census (1950); WU CPB statement (1950); *T&T Age* (October 1950), 50; U.S. Communications Policy Board, *Telecommunications: A program for progress* (Washington, D.C.: U.S. GPO, 1951); Federal Communications Commission (1939–57).

14. On technological momentum, see: Thomas Hughes, "The evolution of large technological systems," in *The social construction of technological systems: New directions in the sociology and history of technology*, ed. Wiebe Bijker, Thomas Hughes, and Trevor Pinch (Cambridge, MA: MIT Press, 1987).

15. U.S. Senate, "Study of the telegraph industry," hearings before a subcommittee of the Committee on Interstate Commerce, 77th congress, 1st session, pursuant to S. Res. 95 (Washington, D.C.: U.S. GPO, 1941); WU, *Annual Report* (1971–47); U.S. Communications Policy Board (1951); WU, "Exhibits originally presented to the president's Communications Policy Board in 1950, . . . extended, revised, or otherwise brought up to date" (1954), WUA, 1993 addendum, series V.

16. American Telephone and Telegraph Company [AT&T], *Events in telecommunications history* (Warren, NJ: AT&T Archives, 1992), 106; *Business Week* [*BW*] (June 8, 1974); Craypo (1979), 299.

17. *BW* (June 8, 1974), 68, 72; WU, *Western Union: From wire to Westar* (New York: WU, 1976).

18. Benjamin S. Rosenthal, *Western Union: The reluctant messenger—a consumer study* ([S.I.: s.n.] 1974), 29–31; *BW* (June 8, 1974). In 1970, President Nixon reorganized the congressionally funded U.S. Post Office Department into a new, separate, government-owned corporation, the U.S. Postal Service. Gerald Cullinan, *The United States Postal Service* (New York: Praeger, 1973), 3–4.

19. McFall (1971), 13; Rosenthal (1974), 1, 10; Philip M. Doyle, "A quarter century of wage gains for telephone and telegraph workers," *Monthly Labor Review* 97:7 (1974).

20. Gerald W. Brock, *The telecommunications industry: The dynamics of market structure* (Cambridge, MA: Harvard University Press, 1981), 280.

21. Jeffrey L. Covell, "New Valley Corporation," *IDCH*, vol. 17, 346–47; Roger W. Rouland, "Curtiss-Wright Corporation," *IDCH*, vol. 10, 260–63; Scott M. Lewis, "Brooke Group Ltd.," *IDCH*, vol. 15, 71–73.

22. AT&T (1992), 169–70; Covell, 345–46; Lewis, 73; W. Bechtel, personal interview (1998). The name "New Valley" evokes WU's first name from 1851, the New York and Mississippi Valley Printing Telegraph Company.

23. Postscript: When my credit card bill came the next month (in the mail), besides the $9.95 charge for the telegram, my credit card company had levied a $5.05 charge because the telegram was somehow classified as a "cash advance." Fifteen more minutes of customer service to have the charge removed (over the telephone), and my Western Union experience finally came to an end— after using the web, the post office, the telephone company, and the Airborne Express courier service.

24. Rosenthal (1974), 1, 11, 13; E. F. Sanger, personal correspondence (1998). The company itself had relocated to Upper Saddle River, New Jersey, in 1969.

25. Robert J. Saunders, Jeremy J. Warford, and Björn Wellenius, *Telecommunications and economic development*, 2nd ed. (Baltimore: Johns Hopkins University Press, 1994); Glenn Collins, "Selling online, delivering on bikes: Low-tech couriers thriving," *New York Times* [*NYT*] (December 24, 1999).

26. *Quicksilver* (USA, 1986); *Double Rush* (USA, 1995); William Gibson, *Virtual light* (New York: Bantam, 1993); Culley (2001). On the "informational city," see: Manuel Castells, *The rise of the network society*, vol. 1 of *The information age: Economy, society, and culture* (Cambridge, MA: Blackwell, 1996); Stephen Graham and Simon Marvin, *Telecommunications and the city: Electronic spaces, urban places* (New York: Routledge, 1996). On bike messenger issues today, see: International Federation of Bike Messenger Organizations (http://www.messengers.org/).

27. Dinitia Smith, "Fast company: Wheel tales of Manhattan's bike messengers," *New York* (January 13, 1986), 38–43; Deborah Frost, "Wild in the streets," *Women's Sports and Fitness* (May 1987), 33–36; Don Cuerdon, "I was a New York City bike messenger," *Bicycling* (March 1990), 74.

28. Smith (1986); Cuerdon (1990); Culley (2001), 290.

29. Culley (2001), 21.

30. Culley (2001), 93–95, 104–5.

31. Zoe Noe, "Get hot!" in *Bad attitude: The processed world anthology*, ed. Chris Carlsson (New York: Verso, 1990), 50–53; Laura McClure, "Bike messengers answer union's call," *The Progressive* 58 (December 1994): 13.

32. Collins (1999); Brad Wieners, "Kozmo's high hopes," *Wired* (February 2000), 70.

33. Associated Press, "Kozmo delivers last rites," *Wired News* (April 12, 2001); Jayson Blair, "Kozmo to end operations; 1,100 laid off," *NYT* (April 12, 2001).

34. Jennifer L. Rich, "Upgrading the last link in the dot-com chain," *NYT* (January 3, 2001).

35. R. Arias, "Kamikaze bikers," *People Weekly* (August 5, 1985), 102–4; Gibson (1993); Culley (2001).

Greg Downey received his bachelor's and master's degrees in computer science from the University of Illinois at Urbana-Champaign, College of Engineering, in 1987 and 1989, respectively. He worked for five years in Chicago as a software developer, first in the advertising industry and then in an "artificial intelligence" academic research laboratory. While working days, he attended evening classes and earned a master's degree in liberal studies from Northwestern University in 1995. Downey subsequently decided to change his career and pursue full-time graduate study toward a doctoral degree in the social and environmental issues surrounding information technology. Admitted to Johns Hopkins University jointly by the Department of History of Science, Medicine, and Technology, and the Department of Geography and Environmental Engineering, Downey was trained in both history of technology and human geography under advisers David Harvey, Stuart Leslie, and Erica Schoenberger, earning his Ph.D. in 2000. After one year at the University of Minnesota Twin Cities Department of Geography as a Woodrow Wilson Postdoctoral Fellow, Downey became an assistant professor at the University of Wisconsin-Madison in both the School of Library and Information Studies and the School of Journalism and Mass Communication. He currently resides in Madison, Wisconsin with his wife, Julie, his son, Henry, his two cats, Coco and Kiwi, and of course his bicycle.

Printed in the United States
by Baker & Taylor Publisher Services